Dams, Development and Environmental Flow in the Damodar River Basin

Md. Mofizul Hoque • Aznarul Islam

Dams, Development and Environmental Flow in the Damodar River Basin

An Integrated Approach

Springer

Md. Mofizul Hoque
Department of Geography
Aliah University
Kolkata, West Bengal, India

Aznarul Islam
Department of Geography
Aliah University
Kolkata, West Bengal, India

ISBN 978-3-032-11160-9 ISBN 978-3-032-11161-6 (eBook)
https://doi.org/10.1007/978-3-032-11161-6

This Springer imprint is published by the registered company Springer Nature Switzerland AG
The registered company address is: Gewerbestrasse 11, 6330 Cham, Switzerland

Thousands of people who were displaced and whose lives and livelihoods were distressed due to dam construction. This work is also devoted to students, academics, and policymakers concerned with the restoration of the environmental flow in the Damodar River Basin.

Foreword

Around the world, rivers and river basins are the primary shapers of landscapes, livelihoods, and resources—and it is fair to say, we are just beginning the journey on how to manage them, collectively for the longer term.

The Damodar—or the so-called *Damuda*, meaning river of sacred water—has a special place on this journey. Once known as the '*Sorrow of Bengal*' due to frequent flooding disasters, also 'full of fire' with its basin of coal and mineral resources, it is now a 'River of agony' with widespread pollution and degradation. The Damodar is also highly significant for the first river valley development corporation (1948), building the first multi-purpose dam (1953), in a newly independent India.

Around 75 years on, this is a good moment to take stock: looking at a wide spread of evidence, bringing together the science of hydrology with new thinking on environmental flows, and the wider experience of sustainable development in practice.

In that light, behind the impressive detail in this book lie three profound challenges in 'joined up thinking': knowledge integration, policy integration, and practical integration.

Challenges of knowledge integration are emerging everywhere, as specialists look beyond their fields and disciplines. We need better ways to connect hydrology and environmental science, with landscape and ecology, agriculture and forestry, rural livelihoods and economic development, social and demographic change, culture and colonial history, and indigenous and grassroots knowledge. There are new opportunities in the digital domain, such as remote sensing and responsive platforms, and new threats from climate change and the increases in flooding, cyclones, droughts, and heatwaves. How to integrate such knowledge is all work in progress at local, national, and global levels.

Challenges of policy and governance integration then follow on from the science: how to design policy around multi-level environmental flows, with multiple stakeholders and multiple objectives? Does this work best at local, district, state, or national level, and how can we get the participation and engagement of all involved? How can governance work when funding is tight, policies are controversial, and the evidence is uncertain?

Practical challenges then follow from this moving picture: should the villagers stay or go in an area of disruption, pollution, and degradation? Should a local enterprise invest with security in the landscape, or migrate to the city like so many others? And overall, could the Damodar Valley Corporation and similar bodies do it all better—in terms of environmental flows and other forms of integration—and what could that mean in practical terms?

This book cannot provide all answers to all questions, but it does provide the evidence base to move forwards. In that sense, it is a crucial contribution to the art and science of living with rivers on one small planet; those living with, on, or around rivers and their landscapes would do well to read it.

Mind-Lab/Laboratory for Urban Collective Intelligence Joe Ravetz
Future Cities Lead, Manchester Urban Institute
University of Manchester
Manchester, UK
25-11-2025

Foreword

I appreciate the invitation from Drs. Md. Mofizul Hoque and Aznarul Islam to write a foreword to the book entitled *Dams, Development and Environmental Flow in the Damodar River Basin: An Integrated Approach*. As they describe, the Damodar River in northeastern India has a history of severe flooding and was called the 'River of Sorrows'. Like elsewhere in the region, the monsoon is the major driver of floods, which primarily occur during the late summer months. Human activities have exacerbated floods through unregulated construction, poor urban planning, and various land-use changes, increasing runoff and accelerating sedimentation in the basin. Anthropogenic actions, including industry, the creation of domestic and agricultural waste, and coal and sand mining, have also impacted water quality, making the river hazardous to use for drinking and agriculture.

After the region experienced centuries of flooding, the Damodar Valley Corporation (DVC) was formed in 1948, modelled after the Tennessee Valley Authority (TVA). With help from one of the leading TVA engineers, the DVC aimed to transform the Damodar from a seasonal river to a perennial river and consistently supply water for irrigation outside of the monsoon months. During the 1950s, four multipurpose dams were built, and the Durgapur Barrage and an associated navigation canal were constructed to moderate flood events and provide water for irrigation year-round. Another dam was completed later in 1978. The authors focus on how DVC dams and barrages altered the flow duration of the river, including the flood and low-flow magnitude and frequency in the quest to establish sufficient flows for sustainable development and next steps for restoration. Although flooding has reduced, long-term siltation of the reservoirs is causing a decline in storage capacity, and thus, in the effectiveness of the multipurpose dams for flood control.

As the authors describe, one of the major social consequences of dam construction is the displacement and resettlement of the villages and agricultural lands where the reservoir covers. In the Damodar River basin, the construction of only two dams resulted in the displacement of nearly 100,000 people, most of whom were members of tribes. Tribal people and other citizens were deprived of access to land, forests, and traditional sources of income, leading to marginalization and vulnerability.

Displaced citizens did not get adequate compensation and resettlement and were left without proper housing, employment opportunities, and access to basic services.

As development and associated industry and mining have progressed, so has pollution, earning the river another moniker: the 'River of Agony'. In most places, the river's water quality is poor due to the large amount of industrial effluent, urbanization, agricultural and domestic waste, and mining. There are dozens of cities and towns and hundreds of coal mines and industries. Pollutants in the water harm drinking water and thus human health, including carcinogenic risk, and aquatic biota, including fish, which can harm human health when consumed. When irrigation water is of poor quality, it decreases crop quality, soil health, agricultural outcomes, and potentially human health when consumed. Excessive pollution also makes it harmful for swimming, bathing, and other water uses. Nearly half of the coal in India is found in the Damodar River basin, and sand mining occurs along most of the river corridor. The authors examine many aspects of water quality parameters spatially and temporally, finding that conditions are worse during the pre-monsoon season when water quantity is low. However, floods can spread contaminants over a larger area. The authors advocate that effluent, contaminants, sewage and runoff need to be better managed, as does the unvegetated land, which is a source of sediment.

Local respondents perceived changes in agricultural development because of the dams and increased irrigation. Cropping intensity and productivity are different from the pre-dam to post-dam periods. The authors applied two methods to assess environmental sustainability, including a rapid impact assessment matrix and a mathematical model of environmental sustainability. Generally, there were more negative components compared to positive components of the Damodar dams for both upstream and downstream villages, suggesting that they were unsustainable.

Our understanding of the interactions between rivers, engineering, and people greatly advances through studies in new areas, leading to improved knowledge. The authors synthesize prior works and enhance knowledge of the role of dams and other human activities in transforming this area, using a variety of sources from the literature. The varied maps, graphs, and tables of different aspects of the dams in the Damodar River Basin provide background information that will assist others in conducting further work on these systems.

Given the vast number of dammed rivers throughout the world, this topic is an area of interest to scientists and citizens. More multi-faceted studies of multiple dams and their benefits and impacts are needed as society faces the legacy impacts of past decisions. Given the need to learn more about this topic, I congratulate the authors on their completion of this book focused on dams in the Damodar River Basin.

University of Florida
Gainesville, FL, USA

Joann Mossa

Preface

Dams-induced development and displacement are the two sides of a coin. On the one hand, dams promote development through the generation of hydro-power, irrigation, and agricultural development. On the other hand, displacement is one of the most common outcomes of dam construction. The Damodar Valley Project is the first multi-purpose river valley project in India for controlling the colossal floods in the river basin and promoting agri-industrial development. This theme has been well explored in previous scholarships. The ways in which the Damodar Valley Corporation helped with flood control and the development of agriculture and industry are well documented. However, the question of environmental flow in the context of dams and development has received less attention in previous works. Thus, this book was conceived with the thought of environmental flow alterations due to the massive structural interventions. Both the quantitative and qualitative flow alterations are equally addressed in this monograph with an eye towards their restoration. This may help not only local policymakers but also people across the world facing similar problems of dam-induced development, industrialization, urbanization, and escalated river pollution. We are hopeful that this work, *Dams, Development and Environmental Flow in the Damodar River Basin: An Integrated Approach*, will unravel the basic concepts of environmental flow alterations, their drivers specific to dams and barrages, and their restoration.

This volume is not just for scholars and researchers; it is for anyone with a passion for rivers, deltas, and the intricate tapestry of human-environment interactions. Whether you are a planner, an educator, a student, or simply a curious observer, we invite you to join us on this journey of discovery.

We are very much thankful to the respondents during the field survey and other stakeholders directly and indirectly associated with the data collection process. We sincerely acknowledge the valuable insights and guidance recieved from Dr. Balai Chandra Das, Barasat Govt. College, Dr. Sandipan Ghosh, Chandrapur College, and Prof. Sanat Kumar Guchhait, The University of Burdwan that shaped this monograph in many ways. We extend our heart-felt thanks to our laboratory members-Dr. Biplab Sarkar, Dr. Sadik Mahammad, Ms. Susmita Ghosh, and Mr. Suman Deb Barman for their active and overwhelming cooperation during field survey, and data processing. We are especially thankful to our students- Mr. Md Hamidur Rahman, Mr. Abujar Rahaman, Mr. Nishan Mallick, Mr. Sukchand Shaik, Mr. Nurul Hassan, Mr. Md. Abdul Kadir Ahamed, Mr. Ibrahim Mondal, Mr. Fahad Hossain Molla, Mr. Sohel Reja, Mr. Rakesh Ali, Mr. Ratnadeep Chanda, Mr. Akash Mondal, Mr. Samim Aktar, Mr. Moktajul Mondal, Ms. Tahamina Nasrin, Ms. Jesmin Sultana, Mr. Raju Saikh and Ms. Nuzhat Fatma for their active support during the field survey, and data collection. We are also equally thankful to our family members for their days of support and sacrifice throughout this entire journey. All concerned persons from Aliah University are duly acknowledged. We also show our deep gratitude to all other persons who promptly responded to our request for study materials and data support. Moreover, we are really thankful to all the local respondents who directly and indirectly helped us during the primary data collection using a questionnaire and instrumental surveys.

We are grateful to ppa Harmen van Paradijs, Applied Sciences, Springer Nature Switzerland AG, Representative with power of attorney, ppa. Christoph Baumann, Springer Nature Switzerland AG, Asset Manager and Aaron Schiller, Editor, Geography, Sustainability Research, Environmental Science, Paleontology, Springer New York for signing the agreement on this project and for providing the opportunity to use their prestigious pages for manuscripts. We are also grateful to the production team for their meticulous efforts to publish the book as correctly as possible.

We hope the outcomes of this volume will flood the thoughts of scholars, faculty, planners, and stakeholders, providing us the appropriate recognition.

Kolkata, West Bengal, India — Md. Mofizul Hoque
Kolkata, West Bengal, India — Aznarul Islam
30 April 2025

Competing Interests The authors have no competing interests to declare that are relevant to the content of this manuscript.

Contents

About the Authors

Md. Mofizul Hoque completed his Master of Science degree in Geography in 2016 and his Ph.D. in 2024 in Geography both from Aliah University, Kolkata, India. He is currently working as a Research Assistant at the Department of Geography, Rampurhat College, Birhum, West Bengal. He has published 13 research articles in various renowned international journals of geography and geoscience and has contributed 4 book chapters to edited volumes. His principal areas of research include environmental flow, river pollution, soil contamination, and water resource management for sustainable development.

Aznarul Islam completed his Master of Science degree in Geography in 2010 from Kalyani University, West Bengal, and his M.Phil. (2013) and Ph.D. (2016) in Geography from the University of Burdwan, West Bengal, India. He is working as an Assistant Professor in the Department of Geography, Aliah University (AU), Kolkata, India, since 2016 and served as the HoD (Officiating), Geography, AU from November 2019 to February 2022. Previously, he was an Assistant Professor (West Bengal Education Service) at the Department of Geography, Barasat Government College, West Bengal (2014–2016). Dr. Islam served as a Guest Faculty/Visiting Faculty at several institutes, including Jadavpur University, Kolkata, and Kazi Nazrul University, Asansol. He was awarded the UGC Junior Research Fellowship in Geography in 2012. Besides, he secured All India Rank 018/188 in Earth Sciences in the CSIR-UGC NET examination in 2011. He has already published more than 90 research papers, including 60 articles in SCI-indexed journals. He is the principal author of the book *Riverbank Erosion in the Bengal Delta*: *An Integrated Perspective*, published in 2024 by Springer. He has co-authored another two books - 1. Palaeochannels and Water Resource Management - A Study in the Western Bengal Delta and 2. The Nadia Rivers - Transition to Transformation both publised by the Springer. He has also edited 10 other books and contributed more than 26 book chapters to edited volumes and conference proceedings. He has presented papers in more than 35 national and international seminars and conferences. He has also successfully guided four Ph.D. theses (three theses under preparation). He was awarded one UGC Minor Research Project. He completed a Major Research Project funded by the ICSSR, Govt. of India, in 2019 as the sole Project Director. In 2023, he received the Department of Science and Technology (Govt. of India)-Science and Engineering Research Board (DST-SERB) State University Research Excellence (SURE) Award as the Principal Investigator. He has been performing the role of editor and reviewer for various journals published by Springer, Elsevier, Taylor & Francis, Wiley, Nature, MDPI, Frontiers, etc. He is a life member of the Foundation of Practicing Geographer (FPG), formerly known as Indian Geographical Foundation (IGF), Kolkata and the National Association of Geographers, India (NAGI), New Delhi. His principal area of research includes exploring river, water, and society through the lens of applied spatial sciences using quantitative techniques and geo-informatics to reduce fluvial and hydrological hazards and enhance community resilience.

Notations and Abbreviations

%Na	Percentage of sodium
ABF	Aquatic base flow
AD	Anno Domini
ADM	Additional District Magistrate
Al	Aluminium
ANOVA	Analysis of variance
As	Arsenic
ASP	Alloy Steel Plant
ATSF	Atchison, Topeka and Santa Fe Railway
BC	Before Christ
BDO	Block Development Office
BE	Biological/ecological
BIS	Bureau of Indian Standards
BMP	Bengal Paper Mill
BOD	Biological oxygen demand
BSP	Bokaro Steel Plant
C	Celsius
Ca^{2+}	Calcium
CBWQI	Comprehensive Bathing Water Quality Index
Cd	Cadmium
CD Block	Community Development Block
CIA	Cumulative impact assessment
CIFRI	Central Inland Fisheries Research Institute
Cl^-	Chloride
CM	Chief minister
CO_3^{2-}	Carbonate
COD	Chemical oxygen demand
CPCB	Central Pollution Control Board
CPI	Comprehensive Pollution Index
CR	Cancerous risk
Cr	Chromium

CT	Census town
Cu	Copper
CV	Coefficient of variation
DCL	Durgapur Chemicals Limited
DEM	Digital elevation model
DO	Dissolved oxygen
DPL	Durgapur Projects Limited
DRB	Damodar River Basin
DRM	Desktop Reserve Model
DSP	Durgapur Steel Plant
DTPS	Durgapur Thermal Power Station
DVC	Damodar Valley Corporation
DVP	Damodar Valley Project
DVRRC	Damodar Valley Reservoir Regulation Committee
E. flow/EF	Environmental flow
EC	Electrical conductivity
ECL	Eastern Coalfield Limited
EFA	Environmental flow assessment
EFN	Environmental flow needs
EFR	Environmental flow requirement
EI	Eutrophication Index
EIA	Environmental impact assessment
EMC	Environmental management class
EO	Economic/operational
ES	Environmental Score
ET	Evapotranspiration
ETP	Effluent treatment plant
EU	European Union
EWR	Ecological water requirements
F^-	Fluoride
FAO	Food and Agriculture Organization
FC	*Fecal coliform*
FDC	Flow duration curve
Fe	Iron
FS	*Fecal streptococci*
ft	Feet
GERD	Grand Ethiopian Renaissance Dam
GHG	Greenhouse gas
GIS	Geographic information system
GW	Groundwater
Ha	Hectare
HCO_3^-	Bicarbonate
HFC	Hindustan Fertilizer Corporation
Hg	Mercury
HM	Heavy metal

HPS	Hydel power station
HQ	Hazard quotient
IDPs	Internally displaced persons
IFIM	Instream flow incremental methodology
IISCO	Indian Iron and Steel Company
INR	Indian Rupee
IUCN	International Union for Conservation of Nature
IWD	Irrigation & Waterways Department
IWQI	Irrigation Water Quality Index
IWRM	Integrated water resources management
K^+	Potassium
KR	Kelly ratio
KW	Kilowatt
LULC	Land use and land cover
LULUCF	Land-use change and forestry
MAF	Mean annual flow
MAR	Magnesium adsorption ratio
MCL	Maximum contaminant level
MCM	Million cubic metre
Mg^{2+}	Magnesium
MGNREGA	Mahatma Gandhi National Rural Employment Guarantee Act
MH	Magnesium hazard
MLD	Millions of litres per day or megalitres per day
mm	Millimetre
Mn	Manganese
MPN	Most probable number
MT	Metric ton
MU	Million units
MW	Megawatts
Na^+	Sodium
NEERI	National Environmental Engineering Research Institute
NH	National highway
NH_4^+	Ammonia
Ni	Nickel
NO_2^-	Nitrite
NO_3^-	Nitrate
OPI	Organic Pollution Index
Pb	Lead
PC	Physical/chemical
pH	Potential of hydrogen
PI	Permeability Index
PO_4^{3-}	Phosphate
POR	Period of record
PS	Potential salinity
RIAM	Rapid impact assessment matrix

RS	Remote sensing
RSBC	Residual sodium bicarbonate
RSC	Residual sodium carbonate
SAR	Sodium adsorption ratio
SC	Social/cultural
SD	Standard deviation
SDG	Sustainable Development Goal
SDO	Sub-Divisional Officer
SO_4^{2-}	Sulfate
SRTM	Shuttle Radar Topography Mission
SSP	Soluble sodium percentage
STP	Sewage treatment plant
STPS	Santaldih Thermal Power Station
TA	Total alkalinity
TC	Total coliforms
TDS	Total dissolved solids
TH	Total hardness
THI	Total Hazard Index
TPD	Tons per day
TSS	Total suspended solids
TVA	Tennessee Valley Authority
US	United States
USA	United States of America
USEPA	United States Environmental Protection Agency
USSL	United States Salinity Laboratory
VEA	Vietnam Environmental Agency
WBPCB	West Bengal Pollution Control Board
WCD	World Commission on Dams
WHO	World Health Organization
WQI	Water Quality Index
Zn	Zinc

Conversion Table

Length

1 km = 1000 m
1 m = 3.28 ft
1 cm = 0.3937 in

Area

1 ha = 2.4711 Acre
1 Acre = 3.025 Bigha
1 Acre = 4047 m^2
1 Acre =100 Satak
1 Bigha = 20 Katha

Volume

1 Cumec = 35.3147 Cusec
1 Acre-ft = 1233.48 m^3
1 ha m = 10,000 m^3

Weight

1 Quintal = 100 kg
1 Quintal = 2.6792 Maund
1 Maund = 40 kg

List of Figures

List of Tables

Chapter 1
Introduction

1.1 Conceptualizing Dams

1.1.1 Concept of Dams

A dam is a manmade or artificial barrier that is installed across a stream, river, or estuary to store or impound water and create a reservoir. It is an anthropogenic structural intervention on a river system or natural stream used to collect water in a reservoir. The upstream side is the barrier side where the water is stored, and the downstream side denotes the opposite side. The Jawa Dam in Jordan is the oldest recognized dam from the year 3000 BC. Dams are usually made from materials such as wood, rock, concrete, earth, or steel. They often include spillways, movable gates, and valves to control water levels and release excess water. They are usually provided with spillway systems to pass the overflow around or through the dam safely. Dams and reservoirs are often used interchangeably, but they are different things. A dam is a physical structure that holds back or retains water, while a reservoir is a body of water created by a dam.

A dam is the foundation for developing and managing water resources in a river basin. The main purpose of dams is to provide water conservation facilities. Dams are built to store water for human consumption, irrigation, industrial use, and aquaculture. They are used to control floods by reducing the peak discharge of floodwater and to generate hydroelectric power by controlling the water flow through them. Besides, they help to improve navigation by increasing the water depth in a river to make it easier for ships and barges to travel. They also provide recreation areas by creating lakes for activities like swimming, boating, and fishing. Moreover, they are also built for more than one or multi-purposes; for example, the Maithon and Panchet dams of Damodar Valley Corporation (DVC), the water of a single reservoir is to be used for different purposes such as controlling floods, providing water

M. M. Hoque, A. Islam, *Dams, Development and Environmental Flow in the Damodar River Basin*, https://doi.org/10.1007/978-3-032-11161-6_1

for domestic, industrial, and irrigation use and mining, generating hydroelectric power, and fishing. However, they have become a focus of environmental concern due to their impact on migratory fish and river ecosystems. In addition, large dams inundate vast areas of land that are home to many people, plants, and animals.

1.1.2 Advantages of Dams

Dams provide socio-economic and environmental benefits, and are useful for addressing various water conservation and crisis management issues. (1) The main advantage of dams is water storage. Water is considered a significant natural resource. The creation of water storage is one of the prime objectives of dam installation. Water storage is the main use of dam-reservoirs. The stored reservoir water helps mitigate the water crisis, especially during dry seasons. It ensures a regular water supply for residential, industrial, and agricultural use. (2) Dams store and supply water for domestic, agricultural and industrial use, which leads to socio-economic development. (3) Controlling floods and mitigating their impact is a major concern of dams. Dams help to control the floodwater flow by storing surplus water, diverting it to other uses, or carefully releasing it. This helps to save lives and properties, but results in displacement. (4) Dams play a significant role in providing water for irrigation. They help to create potential irrigated land and significantly solve the crop water problem, especially during the dry season, which leads to an increase in crop production and a boost in agricultural development. (5) Dams play a crucial role in the creation of fishing grounds and fishing. A breeding ground is also created behind the dam or in the reservoirs. (6) Hydropower has emerged as an effective means of addressing the energy crisis in many regions of India. The most important advantage of hydropower generation is that it is a pollution-free and clean energy source. The Tehri, Srisailam, and Sardar Sarovar dams are some of India's top hydroelectric generation stations. (7) Dams often prove effective in retaining harmful sediments. They can also be useful in disposing of hazardous materials and preventing subsequent pollution. (8) Dams also provide recreational facilities in and around the reservoirs for human beings. People can enjoy boating, skiing, and fishing in the still water in reservoirs and picnic around them. The Maithon and Bhakra Nangal dams in India are popular among tourists as recreational centres.

1.1.3 Disadvantages of Dams

Despite the benefits, dams have significant disadvantages. (1) Displacement is one of the major issues among the other social consequences of dam creation. People are displaced during construction. It is a common issue worldwide. For example, the Akosombo dam in Ghana displaced 80,000 people, while the Gezhouba dam in China displaced over 20,000 people. (2) Although dams can generate pollution-free energy, reservoirs often produce and emit a high percentage of greenhouse gases, mainly methane, with some carbon dioxide due to the decomposition of organic matter submerged in reservoir water. (3) One of the most negative consequences of dam creation is the disruption of local ecosystems by altering the biophysical components. Dams disrupt ecosystems by altering stream flow, changing water temperature, tapping sediments, fragmenting ecosystems, and flooding forests and wetlands. Besides, they fragment rivers and disrupt their natural flow, threatening aquatic animals' existence, especially migratory species. Dams are also responsible for disrupting the biogeochemical cycles of river ecosystems and affecting their function and structure. (4) Dams can significantly disrupt the groundwater table. They often raise groundwater levels in areas surrounding the dam due to impounded water. However, they can also lead to depletion in downstream areas by altering the natural water flow and reducing recharge opportunities. This causes an impact on the groundwater balance in a region that depends on the dam's design and operation. (5) Dams can significantly block the natural progression of water to other countries, states, or regions by physically obstructing the river flow. This can potentially impact water availability downstream and create tensions between different political entities that rely on the same water source. This is particularly relevant when a dam is built on a river that crosses international borders.

1.1.4 Classification of Dams

Broadly, modern dams are to be classified into two types, i.e., concrete dams and embankments. Concrete or masonry dams block streams flowing through narrow canyons, for example, the Nagarjuna Sagar Dam is a masonry dam located in Andhra Pradesh in India. Whereas embankments control streams and rivers flowing through wide valleys, for example, the Tehri Dam is an embankment dam, located in Uttarakhand in India. However, dams are classified into numerous types according to different classifications, which are as follows.

1.1.4.1 Classification Based on Height

Dams are classified into three categories based on their height, which are flows-

1. **Low dams**: A low dam is a dam with a structural height of up to 100 ft.
2. **Medium dams**: Medium dams have a medium height (100–300 ft).
3. **High dams**: High dams have a structural height of greater than 300 ft.

Although there is no universally agreed-upon height limit for each category. The classification is generally used to categorize dams based on their relative height and the engineering considerations involved.

1.1.4.2 Classification Based on Construction Materials

Dams are classified based on the material types used for construction, which is discussed below.

(a) **Earthfill dams:** Earthfill dams, also known as earthen dams, are built from compacted layers of natural materials such as soil, clay, and rock. They are the most common type of dam as they use local materials and are relatively inexpensive to construct. For example, the Tarbela Dam in Pakistan is the world's largest earthfill dam, with a capacity of 142,000,000 m^3. In India, the Banasura Sagar Dam in Kerala is the largest earth dam and the second-largest dam in Asia. Mostly, an earthfill dam is constructed where valleys are wide with flat flanks (abutments). The earthfill dam's height depends on the foundation strength. The earthfill dams are often used to store water for irrigation and domestic use.

(b) **Rockfill dams:** A rockfill dam, also called an earthen dam or terrain dam, is a type of embankment dam made of compacted, granular earth (soil) with a high percentage of large particles or 'rockfill'. The dam also has an impervious zone, which can be made of concrete, masonry, plastic membrane, steel sheet, or wood. Rockfill dams are cost-effective as they use local materials and minimize transportation. They are suitable for construction where appropriate rock can be extracted at or near the dam site and where the foundations will not be subject to settlement due to loading or erosion from any erosion through or below the dam. For example, the Tehri Dam in Uttarakhand is the largest rockfill dam in India, and the Tarbela Dam in Pakistan is the world's largest rockfill dam. The primary purpose of a rockfill dam is to retain water using compacted layers of rock blocks. Primarily, they are used for water resource management, including water storage for irrigation, flood control, and hydroelectric generation, especially in areas where other types of dams may not be feasible due to readily available rock material at the construction site.

(c) **Concrete dams:** Concrete dams are large structures constructed from a mixture of cement, sand, coarse aggregate, and water to retain water. The most common types of concrete dams are gravity, buttress and arch. For example, the Three Gorges Dam in China is the largest concrete dam in the world, and Cheruthoni Dam is the largest concrete gravity dam in India. They are primarily used to store water for numerous purposes such as drinking water supply, irrigation, flood control, hydropower generation, river flow control, improving water quality, sediment control, and recreation, often serving multiple functions at a single site in design and on location.

(d) **Masonry dams:** Masonry dams are dams constructed from stone, brick, or other masonry units joined together with mortar. They can be gravity or arch-gravity dams. They are less common than other types of dams in modern dam construction as they are labour-intensive. For example, Nagarjuna Sagar Dam in India is the largest masonry dam in the world. The primary purpose of a masonry dam is to store a large amount of water in a reservoir that can be used for various purposes such as irrigation, hydropower generation, drinking water supply, flood control, and sometimes recreation by creating a lake behind the dam.

(e) **Steel dams:** A steel dam is a water-retaining structure made from steel instead of common materials like concrete, wood or timber, masonry, or earthworks. Steel dams are a sustainable option for water management, with several advantages over other types of dams. For example, they can be built faster and more cheaply than masonry dams and are statically determined, meaning member strengths and loads can be accurately calculated. For example, the Ashfork-Bainbridge Steel Dam was constructed in the Arizona desert in 1898 to provide locomotive water to the Atchison, Topeka and Santa Fe Railway (ATSF), and the Redridge Steel Dam was constructed in 1901 to provide water to stamp mills in Michigan's Upper Peninsula in the US.

(f) **Timber dams:** A timber dam is a type of dam where the primary structural support is made from wood (timber). They are often built using a structure of logs or timbers to create a barrier against water flow. They are generally used in situations where large-scale concrete or masonry construction is impractical. Examples include timber crib dams and buttress dams with timber cribbing for reinforcement. They were often used in areas with limited access to concrete or where rapid construction was required, especially for small-scale water storage or diversion projects. The Green River Crib Dam is an example of a timber crib dam. It was built on the Green River around 1811 in Guilford, Vermont.

1.1.4.3 Classification Based on Structural Behaviour

Dams are primarily classified into three main categories such as gravity dams, buttress dams, and arch dams based on structural behaviour. Each relies on a different mechanism to resist water pressure: gravity dams use their own weight, arch dams use their curved shape to distribute force, and buttress dams use multiple supporting buttresses to hold back water. They are discussed below-

1. **Gravity dams:** A gravity dam is a structure that uses the foundation's own weight and resistance to retain water. It is the most common type of concrete dam, which uses the weight and friction of concrete to resist water pressure. Gravity dams are large-scale dams, typically made of stone or concrete to hold large amounts of water. They are best suited for rivers in canyons or wide valleys and require a solid bedrock foundation. They are a popular choice for dam projects due to their stability, simple construction, and low-environmental impact. However, they may be limited by a lack of suitable locations and high construction costs, especially in difficult terrain. Some examples of gravity dams include Bhakra Dam, Alcantarilla Dam and Hoover Dam (curved gravity dam). The Bhakra Dam is the highest straight gravity dam in India, and it is about 226 m high and 518 m long.

2. **Buttress dams:** A buttress dam is a type of dam that has watertight walls on the upstream side and triangular-shaped walls on the downstream side. The walls are called buttresses and support the dam at regular intervals. Buttress dams can be straight or curved and are usually made of reinforced concrete. They are a modification of gravity dams, which are concrete or masonry dams that resist overturning and sliding forces by their own weight. Hence, they are a type of gravity dam that uses buttresses to transfer forces towards the dam's foundation. However, they use less concrete than gravity dams but may cost more due to the need for skilled labour, reinforcement, and formwork. They have less stringent foundation requirements than gravity dams. Some examples of buttress dams include Errochty Dam, Sasanagare Dam, Onbara Dam, Grandval Dam, and Latyan Dam, etc.
3. **Arch dams:** Arch dams are concrete or masonry structures that are curved upstream and designed to carry the force of water against them. These dams are generally thin in cross-section and are driven laterally by the forces of reservoir water. They are made from concrete and are usually constructed in narrow, steep-sided valleys with good rock for their foundations. They are thinner than other types of dams, thus they require fewer construction materials and are economical and practical in remote areas. Some examples of arch dams like the Inguri Dam, Chirkei Dam, Vaiont Dam, Ertan Dam, Mauvoisin Dam, El Cajón Dam, and the Swiss Contra Dam, etc. Moreover, the Idukki Arch Dam in Kerala is India's largest arch dam at 168.91 m (554 ft) tall and was the first arch dam in Asia.

1.1.4.4 Classification Based on Hydraulic Design

Dams may be classified into two categories according to hydraulic design, which are discussed below.

1. **Overflow dams:** An overflow dam, also known as a weir, is a small dam used to create a lake to store water for various purposes. It can be used to measure flow, retard flow, and abstract water. Overflow dams are usually part of a gravity or buttress dam. The spillway portion of the dam is lower than the rest, allowing water to flow upward and forward. They allow excess water to pass over them and flow downstream, reaching their intended storage capacity. An example of an overflow dam is a spillway.
2. **Non-overflow dams:** A non-overflow dam is a dam that is not designed to be overtopped. In the case of a non-overflow dam, water is not allowed to flow over the top of the dam. The top of the dam is placed above the highest possible flood level. They are primarily used to store water during high flood events without allowing excess water to spill out. An example of this type of dam would be a large earthfill embankment dam. Nearly all river valley projects are a combination of overflow and non-overflow dams. Examples of non-overflow dams are earth dams and rockfill dams.

1.1.4.5 Classification Based on Rigidity

The most common classification of dams may be classified based on their rigidity, which is as follows-

1. **Rigid dams:** A rigid dam is a dam built from rigid materials like concrete, masonry, steel, or timber. Rigid dams are quite strong. They rarely deflect or deform due to water pressure and other forces. There are several types of rigid dams, including gravity dams, arch dams, steel dams, timber dams, etc.
2. **Non-rigid dams:** Non-rigid dams are constructed from non-rigid materials such as earth and or rockfill. They are less strong compared to rigid dams. A non-rigid dam has relatively considerable settlement and deformation compared to a rigid dam. The most common types of non-rigid dams are earth dams, rockfill dams, and combined earth and rockfill dams.

1.1.4.6 Classification Based on the Function

Dams can be classified into four main types based on their function, which are discussed be low-

1. **Storage dams:** A storage dam is a structure that controls the river flow, especially during the dry season, to continue hydropower generation. They are constructed to retain water when the river flow is high during the monsoon. They can provide water sources or better fish and wildlife habitat. They can also create reservoirs upstream to store water for irrigation and drinking, and to control flooding downstream. The maximum dams are storage dams; the term 'dam' usually denotes a storage dam, as each dam creates a storage behind it. Examples of storage dams are gravity dams, earth dams, rockfill dams, and arch dams.
2. **Diversion dams:** A diversion dam is a dam that diverts river water from its natural course into a canal or other artificial waterway. Generally, diversion dams have a small water storage area upstream and are relatively low in height. Primarily, they are used to divert water for many purposes such as irrigation, hydropower generation, flood control, on-ground or groundwater reservoir creation and storm drains. An example of a diversion dam is a weir or barrage.
3. **Detention dams:** A detention dam is a structure built to control the water flow in a river or stream to hold surface runoff and stream water. They are used to reduce flood damage, control the channel flow rate, and recharge groundwater. Besides, they have minimal impact on the rivers' natural flow, which helps to avoid negative environmental and socio-economic impacts. A detention dam is often called a water-spreading dam or dike and is known as a debris dam due to trapping sediment in the reservoir.
4. **Cofferdams:** A cofferdam is a temporary structure that keeps water and soil away from the excavation area, allowing safe construction in dry environments. They are often used in construction projects that involve working in or near water, such as bridge repairs, marine construction, and oil and gas projects. They

create a dry work environment by pumping water from the enclosed area. This allows workers to carry out construction work safely. There are different types of cofferdams, including enclosed box-type cofferdams and cofferdams that support excavation operations. They can also be designed to be removable, fixed, or mobile. The first recorded use of a cofferdam was in 539 BC by King Cyrus of Persia, who used earthen cofferdams to divert the Euphrates River and capture Babylon. The Romans also used cofferdams to build Trajan's Bridge in 102 AD.

1.2 Dams, Development, and Environmental Flow

A dam is an obstruction that stops the natural flow regime of any channelized water on the earth's surface. A large dam is the most significant and visible construction installed for water resource management, and it has emerged since the twentieth century (Hoque et al., 2022a). In the twentieth century, large dams arose as significant and noticeable structural interventions for water resource management. The installation of large dams had increased tremendously from the 1930s to the 1970s and peaked in the 1970s for socio-economic development (WCD, 2000). In the world, more than 45,000 large dams play a key role in overall societal development, such as providing water for irrigation in agricultural fields, hydroelectricity generation, flood control, water supply for domestic uses, navigation, fishing grounds, and recreation areas. It is estimated that dams across the world serve 30%–40% of irrigated land and generate 19% of the world's electricity (WCD, 2000). The trend of dam construction is high in arid and semi-arid regions (Altinbilek, 2002). This trend across the large rivers is being increased substantially, and therefore, more than half of the world's large rivers are dammed. Consequently, the dam has a great impact on the environmental flow and river ecology by altering the downstream flow patterns, water quality, sediment movement and deposition, habitation, and diversity of aquatic life, and finally the livelihoods of people depending on a healthy river ecosystem (Postel & Richter, 2012).

However, in the context of environmental flow, dams play both negative and positive roles depending on space and time (Dwivedi et al., 2010). For example, the riparian area downstream of a dam may increase, and non-native species may also increase (Zeiringer et al., 2018). Besides, the river loses flow due to the water-holding structure of the dam, which impedes the natural river flow regime and impacts the environment, ecology, aquatic life, and river morphology (Mitra & Singh, 2018). The natural flow of rivers is important for aquatic communities and their ecological integrity because the evolution, adaptation, development, and maintenance of river habitat depend on the continuous flow of water (McManamay & Bevelhimer, 2013). Environmental flow design requires the freshwater quantity, quality, and timing of water flow to improve the natural biogeochemical process and diversify aquatic communities. Therefore, to minimize the negative impacts of dams and sustain the freshwater ecosystem, there is a need to manage the environmental flow (The Brisbane Declaration, 2007). Environmental flow management in the river system can be achieved through decreasing the rising impacts of the hydrological alteration (Tonkin et al., 2014). Hence, environmental flow management requires the establishment and enforcement of river basin planning, industrial reforms, and maintenance of existing infrastructure to promote sustainable development (World Bank, 2008).

1.3 Bibliometric trends in Dams, Development and Environmental Flow Analysis

Dams and development are one of the frequently discussed issues in the contemporary world for their multi-dimensional nature, ranging from ecology to the economy. There are around 30000 documents on dams and development in different languages from 1914 to 2025 based on the Scopus database. However, once the "Environmental flow" is added to the string, the documents count down to 198 as per the Scopus database from 1995 to 2025. This clearly demonstrates that the perspectives of the environmental flow are less explored in the previous scholarships. However, this perspective has become one of the most important issues nowadays because of the fall in both river water volume and quality. The present work is focused on exploring this aspect of the research. We have performed a bibliometric analysis on the spatiotemporal dimension and subject contents using 198 documents. The temporal trend shows that there has been a rise in the document's production since 1995; however, this rise is not at all steady rather more fluctuating, with the highest documents recorded in 2022, followed by a declining trend in recent times (Fig. 1.1A). Regarding the spatiality, the USA exhibits the maximum documents followed by Australia, China, India, and the UK (Fig. 1.1B). This indicates India's stand in the environmental flow analysis in the wake of environmental awareness and achieving the Sustainable Development Goals. Regarding the individual works, Arthington topped the list with five documents. The same metrics re found for Brown and Kingsford. The authors of the present monograph have some important works in the concerned field (Fig. 1.1C). One of the fascinating trends of this bibliometric analysis is to sketch out the interdisciplinary nature of the subject (Fig. 1.1D). Though this topic is greatly focused on the environmental sciences, Earth and environmental sciences, agricultural science, Engineering even in social sciences, the environmental flow analysis in the context of the dams is discussed.

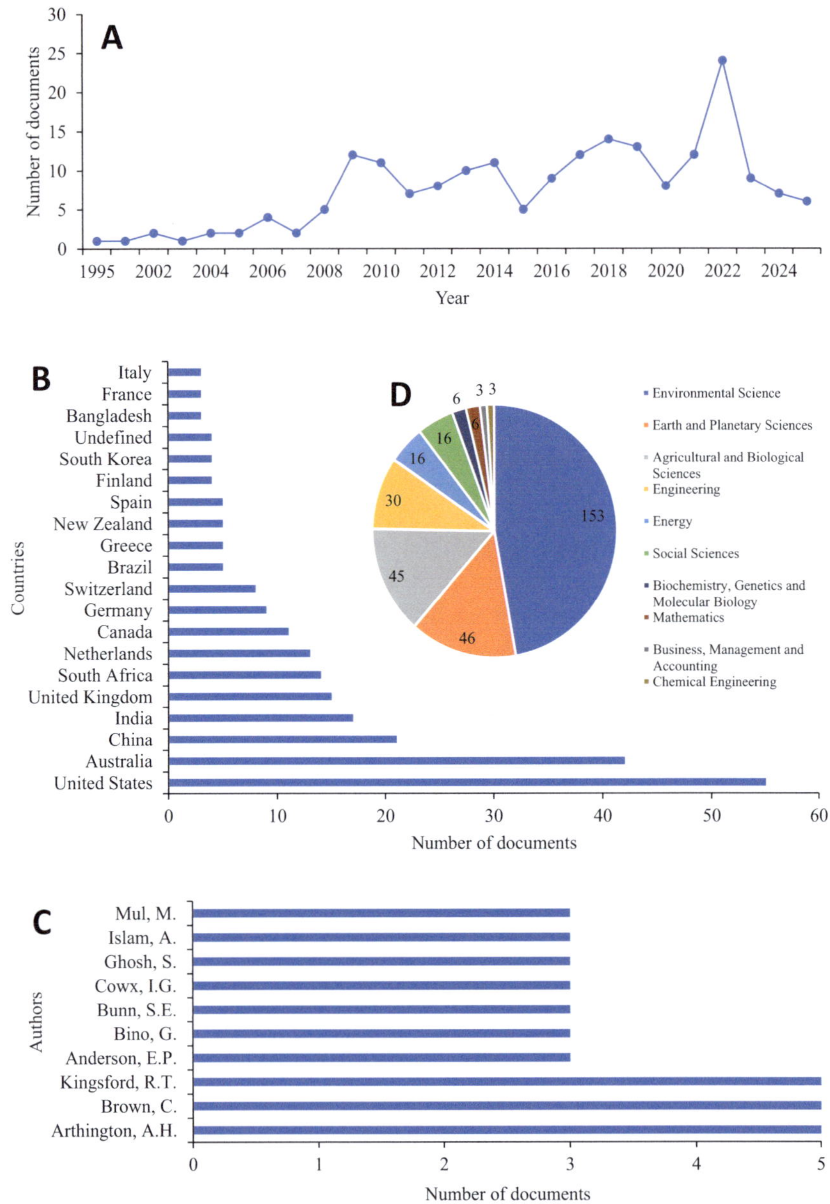

Fig. 1.1 Dams, Development and Environmental Flow in Bibliometrics, (**A**) Temporal growth of the works, (**B**) Spatial growth of the works in different countries, (**C**) Top ten authors working in the concerned field, (**D**) Interdisciplinary nature of the present work

1.4 Approaches to Environmental Flow Measurement and Restoration

Assessment of the environmental flow of a river is essential in the era of the 'Anthropocene' (Poff & Matthews, 2013; Datry et al., 2018). The multiple ecosystem services involved with the environmental flow are significantly degrading in response to the direct and indirect modification of environmental flow by diverse anthropogenic interventions (Ghosh et al., 2022). Therefore, questions are raised on the sustainability of the ecosystem services along with the livelihood of the people and the overall development of society (The Brisbane Declaration, 2007). The concept of environmental flow is widely applied to the fluvial system across the world (Tharme, 2003). It strikes the quantitative and qualitative dimensions of flowing water required for sustaining riverine ecosystem services and socio-economic development (Jain, 2015). The quantitative dimension mainly includes the measurement of river discharge and securing minimum flow for sustaining riverine ecology, while the qualitative dimension focuses on the quality of river water. Environmental protection and sustainable water resource management are very crucial for living organisms and non-living objects. As sustainable water resource management is essential to protect the environment, dams are introduced to regulate the river for water resource management and human water needs (Mitra & Singh, 2018). However, the dam poses several negative impacts on the ecological condition of the river system and its services (Hoque and Islam, 2024).

1.4.1 Measurement of Environmental Flow

Environmental flow refers to the minimum flow of the river required to sustain the ecological stability of the river (Richter et al., 2006). Moreover, it maintains the natural biogeochemical process and habitat suitability of the different aquatic species (Arthington et al., 2010). This flow is also taken into consideration as integrated water resource management (Maran, 2004). Therefore, it can be gauged from different dimensions like the quality, quantity, and duration of river water, and the health of the riverine ecosystem (Tharme, 2003; The Brisbane Declaration, 2007; Chen & Wu, 2019; Sarkar & Islam, 2020). At present, the flowing river water has been excessively used to meet public demand by supplying water to different sectors of the economy (Hoque et al., 2022b, 2024). Consequently, measuring and maintaining the environmental flow has become a major concern among environmentalists (Overton et al., 2014). Therefore, different methods, i.e., flow duration curve, Tennant method, hydraulic rating, and habitat simulation, have been developed. However, environmental flow assessment in the recent decade has focused on the dam-regulated river ecosystem (Tegos et al., 2018). In the last 50 years, various methodologies have been proposed, emphasizing that a minimum flow is needed to mitigate the effects of the dam on the downstream environmental flow (Efstratiadis et al., 2014).

1.4.2 Environmental Flow Restoration

Environmental flow restoration, also known as ecological flow restoration, refers to the practice of reviving and managing the natural water flow in rivers for improving and sustaining the health of aquatic ecosystems and the services they provide. It requires the continuous assessment and maintenance of minimum flow from both the qualitative and quantitative perspectives (Tremblay, 2010). The better conservation and recovery practices of environmental flow include the development of efficient river hydrological modelling based on morphological, hydrological, and ecological information (Arthington et al., 2010). The formulation and execution of effective planning for designing and operating dams can sustain the environmental flow and restore ecological health (King & Brown, 2018). However, the negative consequences of the dam should be minimized (King & Brown, 2018).

1.5 Environmental Flow Studies in the Global Context—North-South Disparity

Environmental flow (EF) is a crucial aspect of the Sustainable Development Goal 6 (SDG 6) (Sood et al., 2017). Studying environmental flow is vital as the world's river flows are increasingly modified by dams, weirs, barrages, and drainage for agriculture, industrial, and domestic supply. Besides, increasing pollution is a significant factor that alters the aquatic systems (Hoque et al., 2025a, b). All over the world, environmental flow studies have gained importance due to growing concern about human-induced changes in water systems and their ecological consequences. Nowadays, it emerges as a vital sustainable water management tool due to increasing pressure on water resources from human activities, climate change and shifting hydrological cycles. It has gradually become increasingly important worldwide, as human activities such as dam construction, water diversion and extraction for agriculture, industry, hydropower generation, and domestic use, as well as land-use change, disrupt natural flow regimes that impact biodiversity and ecosystem services (Hoque et al., 2022a). It helps to balance human water needs with the protection of aquatic biodiversity and ecosystem services. Globally, it emphasizes the necessity to understand, manage, and implement the quantity, quality, and timing of water flows required to sustain aquatic ecosystems and the livelihoods and well-being of the people who depend on them (The Brisbane Declaration, 2007). More such studies on freshwater systems are crucial due to the increasing pressures on these resources from population growth, climate change, and intensive human activities (Hoque et al., 2023). It is essential for understanding the relationship between freshwater ecosystems and water flow systems.

Globally, there is a clear distinction in environmental flow studies between countries in the Global North (mostly developed countries in the Northern Hemisphere, especially in Europe and North America) and the Global South (mostly developing and underdeveloped countries in Africa, Latin America, and Asia). In developed countries, there is a greater awareness and legislation regarding environmental flow requirements, such as the United Nations (UN) Watercourses Convention and the European Union (EU) Water Framework Directive. In developing countries, environmental flow studies may be less extensive, but they are vital to balancing development goals with conservation. Environmental flow studies in northern regions mainly focus on maintaining the aquatic ecosystem's health, preventing pollution, and managing hydropower consequences and migratory fish routes (e.g., salmon). These studies integrate environmental flows into larger water management frameworks with regulatory systems that are well-established and enforced. However, in the southern regions, these studies focus on ensuring environmental sustainability and addressing the water requirements for human livelihoods.

1.6 Environmental Flow Studies in the Humid Tropics

Environmental flow studies in the humid tropics primarily focus on rivers in regions such as the Amazon rainforest, Southeast Asian River systems, and parts of Australia (e.g., the Daintree River). They assess the impact of altered flow regimes on aquatic ecosystems due to various factors such as dam construction, land-use changes, and climate variability. Some examples of environmental flow studies in the humid tropics such as the impact assessment of hydropower dams on fish migration and floodplain dynamics in the Amazon River Basin (Chen et al., 2023); the effects of agricultural practices, deforestation, mining, and land degradation on river flow and water quality in the Southeast Asian Rivers (Hairan et al., 2021); the impacts of climate change on river flows and their implications for aquatic biodiversity in the Central American Rivers; and investigation on changing water flow due to dam operations impact on fish spawning, wetland ecology, and the communities' livelihoods rely on the river's ecosystem in the Mekong River Basin (Yoshida et al., 2020; Sor et al., 2023; Cowx et al., 2024).

Several methods, such as hydrological monitoring, ecological surveys, and habitat and hydrodynamic modelling, have been used for environmental flow studies in the humid tropics. They focus on determining or assessing the level of water flow required in rivers to maintain healthy aquatic ecosystems. The humid tropics are a bioclimatic region characterized by consistently high temperatures and rainfall. These regions have unique hydrological processes, such as rapid runoff and high evapotranspiration, that significantly impact river flow. Besides, these regions have some challenges, such as high biodiversity and complex hydrological dynamics due to heavy precipitation and dense vegetation cover, which make it difficult to determine the specific flow requirements of different species. Furthermore, rapid land-use changes and water extraction can complicate environmental flow assessments. Thus, specific considerations are required when determining environmental flow needs. Some important aspects that need to be considered for determining the environmental flow in the humid tropics such as high variability in flow (high flow in rainy and low flow in dry seasons) or rapid flow fluctuations in rivers due to intense rainfall events; impacts of land-use change as deforestation and agricultural expansion alter river flow regimes, often increase peak flows and reduce base flows, which can negatively affect aquatic ecosystems; species-specific requirements as the different aquatic organisms have specific needs related to water depth, flow velocity, and habitat availability depending on life stages (e.g., spawning, migration, and feeding); and importance of floodplains as several rivers in humid tropical depend on floodplain inundation to maintain biodiversity.

1.7 Environmental Flow Studies in South East Asia

Southeast Asia is a riverine region and the world's most dynamic area, characterized by a growing population, a high dependence on water resources, and a highly variable climate across its different parts (Hairan et al., 2021). Rivers like the Indus, Ganges, Brahmaputra, and Mekong serve as lifelines for millions of people in Southeast Asia who are struggling to reach the sea. The South Asian rivers' flow has been diverted for irrigation, industrial, and domestic purposes, and their flow has been regulated by dams, barrages, and embankments. Pollution from industries and households poses a major issue for South Asian rivers, as effluents are discharged into the river system with little or no treatment. In recent decades, political developments regarding water disputes have also risen between neighbouring countries and states (Gopal, 2013). Consequently, South Asian rivers have received substantial attention from researchers, water resource managers, various stakeholders, and politicians. Environmental flow (e-flow) studies in Southeast Asia primarily focus on understanding and maintaining water flows in rivers and wetlands to support ecosystem health, biodiversity, and human livelihoods. This is an important aspect of balancing development with ecosystem conservation. These studies are particularly vital in a region experiencing rapid economic growth, urbanization, and significant changes in its river systems due to developmental projects such as dams, irrigation systems, and industrial development. However, they face challenges due to limited information, complex governance, and competing demands for water resources.

1.8 Research Gaps

There are ample numbers of research works addressing the dam-induced socio-economic progress across the world (WCD, 2000; Shah & Kumar, 2008; Tortajada et al., 2012; Tilt, 2014; Aeria, 2016; Shi et al., 2019). Many scholastic works have also addressed a similar dimension concerning the DVC dams. Particularly, DVC dams have been explored emphasizing flood and drought mitigation (Ghosh & Mistri, 2015), changes in hydro-geomorphology (Islam et al., 2023; Ghosh et al., 2022), river ecology (Banerjee & Ghosh, 2016), irrigation and drinking water supply (Kumar & Singh, 2022), water pollution (Seal et al., 2022), and dams-induced displacement and habitations (Negi & Azeez, 2022). However, the dimensions of the dam and the environmental flow of DVC have remained unexplored. Therefore, this study has attempted to explore the ecological dimension of the environmental flow influenced by DVC. Moreover, the environmental flow is strongly linked with the socio-economic development. Hence, addressing an integrated outlook among the dams, environmental flow, and socio-economic development has been considered as another important research gap that can be meticulously addressed in this study. Furthermore, the absence of policy frameworks for environmental flow restoration has also been observed as another unexplored dimension of the research.

1.9 The Need for the Present Study

In India, the environmental flow has been degraded day by day due to the emergence of hundreds of large dams constructed on rivers to meet the increasing demand of the people. Now, environmental flow needs to be maintained; therefore, recommendations are made by various authorities (Courts, Tribunals, Expert Appraisal Committee of the Ministry of Environment and Forests, India) for releasing the threshold discharge in favour of healthy environmental flows. There have also so many controversies about whether to remove the large dam or not in the last few decades due to having multiple negative consequences. Being a developing country, there are also many problems that have arisen in the last few years in India.

In the study area, dams have also been constructed mainly to control floods, for irrigation and hydroelectricity generation purposes under the scheme of DVC. It was established on July 7, 1948, by an Act of the Constituent Assembly of India (Act No. XIV of 1948), and it is the first multipurpose river valley project of independent India. Though DVC was planned for the well-being of society, the dams created many hindrances like sedimentation on the river bed, the transformation of the river bed, reduction of fish habitat, degradation of water quality, interruption of the natural flow of water and displacement of human habitation, etc. Moreover, after the construction of the DVC dams and barrages, the fluvial modification of the active channel area is evident in the downstream channel, and anthropogenic intervention is escalating in the active river bed, hampering the free flow of the river. The holistic benefits of water resource management are fruitfully enjoyed by the mass population of the Damodar Floodplain without compromising the future of the sustainable fluvial environment and ecosystem functionality. The dams were once considered the pillar of economic development, but now in the study area, the dams are unable to continue the direction of sustainable development strategies to keep pace with the growing demand of population, cities, industries, and agriculture. Unfortunately, the sustainability of the fluvial environment is not getting the right attention from planners and developers. At present, the database shows that the Damodar River is heavily polluted upstream of the Durgapur Barrage, and the riparian area of the river is reduced mainly due to in-channel sedimentation, settlement, agriculture, and sand mining (De et al., 1980, 1985; De, 1993).

The main flow of the river is now maintained, but it is checked within the narrow range of active channels, not sufficient for environmental flow requirements. Alongside the peak or bankfull flow, regulated by dams, persists for 3–4 days only to save the low-lying downstream area of the Damodar during the monsoon. Giving importance to the benefits of dams (especially to humans) is needed to save the river ecosystem of the Damodar from the point of view of environmental ethics. Thus, the study placed more attention on current environmental problems observed in the river due to the DVC project and other structural interventions in the form of check dams, barrages, weirs, and bridges. The present study aims to investigate the major impacts of dams on environmental flow, riverine ecology and societal developments, and also to find out the possible solution to overcome those problems related to the dams in the Damodar River basin.

The Damodar River, which triggered several devastating floods damaging the lives and livelihoods of thousands of people of the DRB, was once called the '*Sorrow of Bengal*' (Bhattacharyya, 2011). Now, the Damodar River is often referred to as the 'River of agony' due to its elevated pollution and riverine degradation caused by massive human interventions (Patra, 2008). The mega-scale interventions broadly encompass the construction of DVC and associated urban-industrial growth, coupled with changes in land use and land cover and intensive agricultural practice. In the Damodar River basin, five large dams (Tilaiya, Konar, Maithon and Panchet under the control of DVC and Tenughat under the control of the Jharkhand Government) and one barrage (Durgapur Barrage under the control of DVC) were installed for moderating the colossal flood events, bringing the socio-economic development of the local people. Though DVC benefited society and the economy in the short run, the damming of the river and other basin-scale development projects vehemently altered the river regime and water quality, especially in the post-dam period, having repercussions on the long-term economic growth. The DRB exhibits several crucial issues for economic development and environmental flows. Thus, the study area has been selected to assess the river's environmental flow in the context of DVC dams and development to find probable pathways to restore the quantitative and qualitative dimensions of the environmental flow of the Damodar River for sustainable development.

1.10 Objectives and Hypothesis of the Study

The present study has a particular focus on the role of DVC in eco-geomorphological behavioural changes and socio-economic development. Hence, the major objectives of the present research are mentioned as follows.

1. To measure the quantitative dimension of the environmental flow of the Damodar River,

2. To characterize the qualitative dimensions of the environmental flow from the perspectives of irrigation suitability, drinking water quality, riverine ecology and biota,
3. To integrate the socio-economic development and environmental flow of the Damodar River and
4. To figure out probable pathways for environmental flow management.

Taking the research problem, objectives, and primary and secondary information, one central hypothesis is developed as mentioned below.

Dams under Damodar Valley Corporation, though play a significant role in developing the regional economy, interrupt the environmental flow for aquatic organisms and human health.

Therefore, to fulfil the major objectives of the study and to establish the hypothesis, the next chapters are systematically organized. The Chap. 2 is thus concerned with the nature and chracterizations of hydrological hazards (floods and droughts) that set the tune why this environmental flow measurement and restoration is essential in the era of climate change and 'Anthropocene' effects.

References

Aeria, A. (2016). Economic development via dam building: The role of the state government in the Sarawak corridor of renewable energy and the impact on environment and local communities. *Southeast Asian Studies, 5*(3), 373–412. https://doi.org/10.20495/seas.5.3_373

Altinbilek, D. (2002). The role of dams in development. *Water Science and Technology, 45*(8), 169–180.

Arthington, Á. H., Naiman, R. J., Mcclain, M. E., & Nilsson, C. (2010). Preserving the biodiversity and ecological services of rivers: New challenges and research opportunities. *Freshwater biology, 55*(1), 1–16. https://doi.org/10.1111/j.1365-2427.2009.02340.x

Banerjee, R., & Ghosh, A. R. (2016). Identification and quantification of aquatic macrophytes with ecological indices in the Damodar River. *International Journal of Scientific Research in Science, Engineering and Technology, 2*(6), 275–281.

Bhattacharyya, K. (2011). *The Lower Damodar River, India: Understanding the human role in changing fluvial environment*. Springer.

Chen, A., & Wu, M. (2019). Managing for sustainability: The development of environmental flows implementation in China. *Water, 11*(3), 433. https://doi.org/10.3390/w11030433

Chen, Q., Li, Q., Lin, Y., Zhang, J., Xia, J., Ni, J., et al. (2023). River damming impacts on fish habitat and associated conservation measures. *Reviews of Geophysics, 61*(4), e2023RG000819. https://doi.org/10.1029/2023RG000819

Cowx, I. G., Vu, A. V., Hogan, Z., Mallen-Cooper, M., Baumgartner, L. J., Lai, T. Q., et al. (2024). Understanding the threats to fish migration: Applying the global swimways concept to the lower Mekong. *Reviews in Fisheries Science & Aquaculture*, 1–29. https://doi.org/10.1080/23308249.2024.2401018

Datry, T., Boulton, A. J., Bonada, N., Fritz, K., Leigh, C., Sauquet, E., et al. (2018). Flow intermittence and ecosystem services in rivers of the Anthropocene. *Journal of Applied Ecology, 55*(1), 353–364. https://doi.org/10.1111/1365-2664.12941

De, A.K. (1993). Growth of water pollution in Asansol—Durgapur industrial belt and its impact on the water quality of the Damodar River. *Environmental Research*. MOEF.

De, A. K., Sen, A. K., & Modak, D. P. (1980). Some industrial effluents in Durgapur and their impact on the Damodar River. *Environment International, 4*(2), 101–105.

De, A. K., Sen, A. K., Karim, M. R., Irgolic, K. J., Chakraborty, D., & Stockton, R. A. (1985). Pollution profile of damodar river sediment in Raniganj-Durgapur industrial belt, West Bengal, India. *Environment International, 11*(5), 453–458.

Dwivedi, V. K., Gupta, S. K., & Pandey, S. N. (2010). A study of environmental impact due to construction and operation of dam. In *National Conference on Eco friendly Manufacturing for Sustainable Development* (Vol. 31, pp. 1–6).

Efstratiadis, A., Tegos, A., Varveris, A., & Koutsoyiannis, D. (2014). Assessment of environmental flows under limited data availability: Case study of the Acheloos River, Greece. *Hydrological Sciences Journal, 59*(3–4), 731–750.

Ghosh, S., & Mistri, B. (2015). Geographic concerns on flood climate and flood hydrology in monsoon-dominated Damodar river basin, Eastern India. *Geography Journal, 2015.*

Ghosh, S., Hoque, M. M., Saha, U. D., & Islam, A. (2022). Assessment of dam-induced changes in ecogeomorphological behaviour and fluvial functionality in the Damodar River, West Bengal, India. *AQUA—Water Infrastructure, Ecosystems and Society, 71*(6), 722–750. https://doi.org/10.2166/aqua.2022.003

Gopal, B. R. I. J. (2013). Environmental flows: The South Asian experience. In *Environmental flow–An introduction for water resource managers* (pp. 183–195).

Hairan, M. H., Jamil, N. R., Looi, L. J., & Azmai, M. N. A. (2021). The assessment of environmental flow status in Southeast Asian Rivers: A review. *Journal of Cleaner Production, 295*, 126411. https://doi.org/10.1016/j.jclepro.2021.126411

Hoque, M. M., & Islam, A. (2024). Spatio-temporal dynamics of ecological, bacteriological, and overall water quality of the Damodar River, India. *Environmental Science and Pollution Research, 31*(12), 18465–18484. https://doi.org/10.1007/s11356-024-32185-5

Hoque, M. M., Islam, A., & Ghosh, S. (2022a). Environmental flow in the context of dams and development with special reference to the Damodar Valley Project, India: a review. *Sustainable Water Resources Management, 8*(3), 62. https://doi.org/10.1007/s40899-022-00646-9

Hoque, M. M., Islam, A., & Mahammad, S. (2022b). Assessing the surface and bottom river water quality for drinking purpose and human health risk analysis: a study of Damodar River, India. *Arabian Journal of Geosciences, 15*(23), 1734. https://doi.org/10.1007/s12517-022-10987-6

Hoque, M. M., Islam, A., Islam, A. R. M. T., Pal, S. C., Mahammad, S., & Alam, E. (2023). Assessment of soil heavy metal pollution and associated ecological risk of agriculture dominated mid-channel bars in a subtropical river basin. *Scientific Reports, 13*(1), 11104. https://doi.org/10.1038/s41598-023-38058-0

Hoque, M. M., Islam, A., Sarkar, B., & Saha, U. D. (2024). Assessing the surface and bottom river water quality for irrigation: A study of Damodar River, India. *International Journal of Energy and Water Resources, 8*(4), 545–562. https://doi.org/10.1007/s42108-022-00206-z

Hoque, M. M., Islam, A., Pal, S. C., Sarkar, B., & Hossain, Z. M. (2025a). Assessment of fish habitat suitability in the wake of elevated pollution of a tropical river in India. *Environmental Science and Pollution Research, 32*(3), 1208–1222. https://doi.org/10.1007/s11356-024-35685-6

Hoque, M. M., Pal, S. C., Mandal, S., & Islam, A. (2025b). Assessment of mollusc habitat suitability of an anthropogenically altered tropical river in India. *Environmental Monitoring and Assessment, 197*(10), 1–18. https://doi.org/10.1007/s10661-025-14593-3

Islam, A., Sardar, N., Mohinuddin, S., Hoque, M. M., Sengupta, S., Das, B. C., ... & Sengupta, B. (2023). Quasi-equilibrium channel metamorphosis in planform of a subtropical river in India in post-dam period. *Catena, 221*, 106793. https://doi.org/10.1016/j.catena.2022.106793

Jain, S. K. (2015). Assessment of environmental flow requirements for hydropower projects in India. *Current Science, 108*(10), 1815–1825.

King, J. M., & Brown, C. (2018). Environmental flow assessments are not realizing their potential as an aid to basin planning. *Frontiers in Environmental Science, 6*(113), 1–8.

Kumar, P., & Singh, A. K. (2022). Hydrogeochemistry and quality assessment of surface and sub-surface water resources in Raniganj coalfield area, Damodar Valley, India. *International*

Journal of Environmental Analytical Chemistry, 102(19), 8346–8369. https://doi.org/10.1080/03067319.2020.1849653
Maran, S. (2004). Environmental flows and integrated water resource management: The Vomano River case study. *Assessment and provision of environmental flows in mediterranean watercourses*. IUCN.uicnmed.org
McManamay, R. A., & Bevelhimer, M. S. (2013). *A Holistic framework for environmental flows determination in hydropowercontexts*. Oak Ridge National Laboratory. http://www.osti gov/bridge
Mitra, S., & Singh, A. (2018). Assessment of environmental flow requirements of Damodar river basins by using flow duration indices method—A case study. *International Journal of Hydrology, 2*(3), 281–283.
Negi, D. P., & Azeez, E. P. (2022). Impacts of development induced displacement on the tribal communities of India: An integrative review. *Asia-Pacific Social Science Review, 22*(2), 50–62.
Overton, I. C., Smith, D. M., Dalton, J., Barchiesi, S., Acreman, M. C., Stromberg, J. C., & Kirby, J. M. (2014). Implementing environmental flows in integrated water resources management and the ecosystem approach. *Hydrological Sciences Journal, 59*(3–4), 860–877.
Patra, S. (2008). *Pollution profile of Damodar River in a particular stretch*. Unpublished Report of Master of Technology in Conservation of Rivers and Lakes, Alternate Hydro Energy Centre, Indian Institute of Technology Roorkee. Retrieved February 15, 2022, from http://shodhbhagirathi.iitr.ac.in:8081/jspui/bitstream/123456789/12088/1/HYDG13946.pdf
Poff, N. L., & Matthews, J. H. (2013). Environmental flows in the Anthropocence: Past progress and future prospects. *Current Opinion in Environmental Sustainability, 5*(6), 667–675. https://doi.org/10.1016/j.cosust.2013.11.006
Postel, S., & Richter, B. (2012). *Rivers for life: Managing water for people and nature*. Island Press.
Richter, B. D., Warner, A. T., Meyer, J. L., & Lutz, K. (2006). A collaborative and adaptive process for developing environmental flow recommendations. *River Research and Applications, 22*(3), 297–318.
Sarkar, B., & Islam, A. (2020). Drivers of water pollution and evaluating its ecological stress with special reference to macrovertebrates (fish community structure): A case of Churni River, India. *Environmental Monitoring and Assessment, 192*(1), 1–31. https://doi.org/10.1007/s11356-021-17719-5
Seal, K., Chaudhuri, H., Pal, S., Srivastava, R. R., & Soldatova, E. (2022). A study on water pollution scenario of the Damodar river basin, India: Assessment of potential health risk using long term database (1980–2019) and statistical analysis. *Environmental Science and Pollution Research, 29*(35), 53320–53352. https://doi.org/10.1007/s11356-022-19402-9
Shah, Z., & Kumar, M. D. (2008). In the midst of the large dam controversy: Objectives, criteria for assessing large water storages in the developing world. *Water Resources Management, 22*, 1799–1824. https://doi.org/10.1007/s11269-008-9254-8
Shi, H., Chen, J., Liu, S., & Sivakumar, B. (2019). The role of large dams in promoting economic development under the pressure of population growth. *Sustainability, 11*(10), 2965. https://doi.org/10.3390/su11102965
Sood, A., Smakhtin, V., Eriyagama, N., Villholth, K. G., Liyanage, N., Wada, Y., et al. (2017). *Global environmental flow information for the sustainable development goals* (Vol. 168). International Water Management Institute (IWMI). https://doi.org/10.5337/2017.201
Sor, R., Ngor, P. B., Lek, S., Chann, K., Khoeun, R., Chandra, S., et al. (2023). Fish biodiversity declines with dam development in the Lower Mekong Basin. *Scientific Reports, 13*(1), 8571. https://doi.org/10.1038/s41598-023-35665-9
Tegos, A., Schlüter, W., Gibbons, N., Katselis, Y., & Efstratiadis, A. (2018). Assessment of environmental flows from complexity to parsimony—Lessons from Lesotho. *Water, 10*(10), 1–17. https://doi.org/10.3390/w10101293
Tharme, R. E. (2003). A global perspective on environmental flow assessment: Emerging trends in the development and application of environmental flow methodologies for rivers. *River Research and Applications, 19*(5–6), 397–441.

The Brisbane Declaration. (2007). Environmental flows are essential for freshwater ecosystem health and human well-being. In *The 10th International Riversymposium and International Environmental Flows Conference, Brisbane, Australia* (pp. 3–6).

Tilt, B. (2014). *Dams and development in China: The moral economy of water and power*. Columbia University Press.

Tonkin, J. D., Jähnig, S. C., & Haase, P. (2014). The rise of riverine flow-ecology and environmental flow research. *Environmental Processes, 1*, 323–330.

Tortajada, C., Altinbilek, D., & Biswas, A. K. (2012). *Impacts of large dams: A global assessment*. Springer.

Tremblay, H. (2010). The emergence of environmental flow protection in Québec Law. *Les Cahiers de droit, 51*(3), 801–825.

WCD. (2000). *Dams and development: A new framework for decisionmaking*. Earthscan Publications Ltd.

World Bank. (2008). *Mainstreaming environmental flow requirements into water resources investments and policy reforms* (Vol. 1). DRAFT, Main Report. Environment Department Energy, Transport and Water Department.

Yoshida, Y., Lee, H. S., Trung, B. H., Tran, H. D., Lall, M. K., Kakar, K., & Xuan, T. D. (2020). Impacts of mainstream hydropower dams on fisheries and agriculture in lower Mekong Basin. *Sustainability, 12*(6), 2408. https://doi.org/10.3390/su12062408

Zeiringer, B., Seliger, C., Greimel, F., & Schmutz, S. (2018). River hydrology, flow alteration, and environmental flow. In *Riverine ecosystem management: Science for governing towards a sustainable future* (pp. 67–89).

Chapter 2
Geographical Personality and Hydrological Hazards in the Damodar River Basin

2.1 Geographic Personality of the Damodar River Basin

The Damodar River is one of the most important rivers in South Bengal (Hoque et al., 2022b). Once, it was known as the *'Sorrow of Bengal'* because of several devastating floods that occurred frequently in the basin area due to high rainfall (<1300 mm/year) (Ghosh et al., 2022). But now, the river has turned into the '*River of agony*' from the environmental point of view due to environmental degradation (Patra, 2008; Bhattacharyya, 2011). It is also called '*Damuda*' which means river of sacred water. Locally, the term 'Damodar' stands as 'full of fire' because it passes through a coal-rich region of the Gondwana formation (Bhattacharyya, 2011). It is a rain-fed river influenced by south-west monsoonal regimes (Hoque et al., 2023). The rain-fed Damodar River rises from the Khamarpet Hill near Chandwa village in Latehar district, approximately at 23°37′ N and 84°41′ E (Bhattacharyya, 2011; Ghosh, 2014). The total length of this river is about 592 km and the average discharge rate is 296 m^3/s measured at Rhondia weir (Hoque et al., 2024, 2025a). It is one of the most polluted rivers in India (Banerjee et al., 2003) as the basin is home to heavy industrial, mining activity, and populated regions (Dutta et al., 2021). The funnel-shaped Damodar River Basin (DRB) is a sub-basin of the river Ganga, covering an area of about 23,370.98 km^2 in the states of Jharkhand (73.7%) and West Bengal (26.3%) in India (Majumder et al., 2010; Islam et al., 2023). The river and its basin area have a great geographic personality in various aspects, such as physio-cultural diversity, coupled with many natural and anthropogenic hazards described below in detail.

Supplementary Information The online version contains supplementary material available at https://doi.org/10.1007/978-3-032-11161-6_2.

M. M. Hoque, A. Islam, *Dams, Development and Environmental Flow in the Damodar River Basin*, https://doi.org/10.1007/978-3-032-11161-6_2

2.1.1 *Physical Aspects*

2.1.1.1 Geology

Different rock types with different geological periods are found in the Damodar River Basin. The geology of the DRB varies from the lower basin to the middle and upper basin (Fig. 2.1). The lower and middle basins are filled by newer and older alluvium in recent geological periods. The older alluvium soils are found over the parent rock during the Tertiary age (Tiwary & Dhar, 1994). The middle basin covers laterites, lateritic soils, and bauxite, which are the formations of the Cainozoic Tertiary period. Late tertiary gravels are also found in the middle basin. More than half the area of the Damodar River Basin is covered by Archean formations (Singh, 1997). The rock type of granite-gneiss in the Archean formation is found in the upper part of the basin (DVC, 1992). The upper and lower Gondwana formation is found in north Karanpura, Jharia, and Raniganj coal fields over the Archean basement in the upper and middle basins. The geological formation of the upper Gondwana systems is sandstone and red clay over the lower Gondwana (DVC, 1992; Geological Map of Jharkhand, 2015). The lower Gondwana systems over the Archean basement are carbonaceous shale, sandstone, and coal seams (Geological Map of Jharkhand, 2015). In the Gondwana period, there had been repeated cycles of erosion and deposition that occurred as per sedimentation evidence from

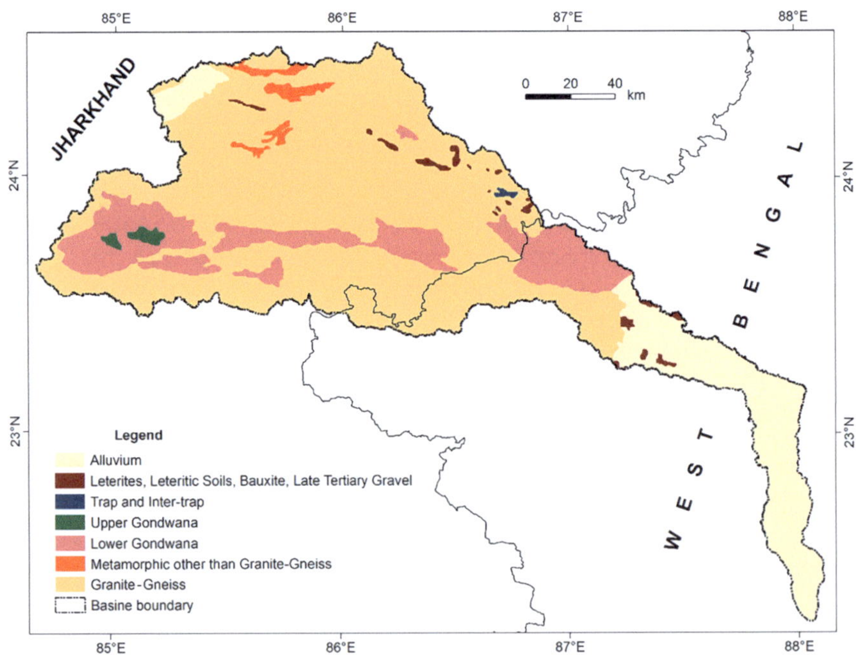

Fig. 2.1 General geology of the Damodar River Basin area based on Bhattacharyya (2011)

Table 2.1 Stratigraphic succession of Gondwana sediments in Damodar Valley

Age		Group	Formation
Lower Cretaceous		Upper Gondwana	Lamprophyre and Dolerite Intrusive
Jurassic	Upper		Non-deposition
	Middle		
	Lower		
Triassic	Upper Rhaetic		Suptra-Panchet Formation
	Middle Noric		Infra-Norian Erosional Surface
	Lower Carnic		Panchet Formation
Permian	Upper	Lower Gondwana	Raniganj Formation
			Barren Measure Formation
			Barakar Formation
			Kaharbari Formation
			Talchir Formation
Precambrian Gneissic Basement of Chotanagpur Granite-Gneiss Complex			

Source: Banerjee (2013)

different formations of Gondwana (Table 2.1) (Mondal et al., 2018). Trap and inter-trap formation topography are also found in the northern edge of the Raniganj coal field in the upper basin, which is formed by trap and inter-trappean beds of basalt, clay stone, silt, stone tuff with trap dykes. The trap and inter-trap formation topography was formed in the upper Cretaceous to lower Eocene periods (Table 2.2).

The Damodar River flows over the Gondwana faulted trough and the Archean granite-gneissic surface in the upper part (Fig. 2.2). In the middle reach, the river flows over the sandstone-rich Gondwana sedimentaries near the Raniganj coalfield (Bhattacharyya, 2011).

2.1.1.2 Physiography and Relief

Physiographically, various distinct features are found in the upper, middle, and lower parts of the basin (Fig. 2.2). The basin can be divided into four physiographical regions, such as (i) Eastern low alluvial plains, (ii) Central rolling plains, (iii) Central uplands, and (iv) Western plateau and plateau fringe (referred to as Rarh plateau in West Bengal) (Bhattacharyya, 2011). Furthermore, it can be divided into four micro-geomorphic units, such as (1) Hills, (2) Plateaus, (3) Degradational and dissected uplands, and (4) Aggradational flat plains based on Bhattacharyya (2011).

(a) **Hills**

Parasnath Hill is the highest peak in the basin (Prasad & Giri, 2009), located towards the eastern end of the Chotanagpur Plateau in the Giridih district with an elevation of 1365 m (Fig. 2.2). The Gondwana formations are observed in the Parasnath Hill. It is formed largely by the Archaean granite-gneiss and meta-sedimentary rock. Hilly areas are found over the Gondwana trough region of the upper basin (Singh, 1997). The hilly regions are generally found along

Table 2.2 The geological succession of the Damodar Valley region

Geological formations	Geological period
Alluvial	Recent
Laterites, lateritic soils, and bauxite	Tertiary
Gravels	Late Tertiary
Trap and Inter-Trap	Upper Cretaceous to Lower Eocene
Dubrajpur Formation	Upper Gondwana
Panchet	
Raniganj	
Barren Measures	Lower Gondwana
Barakar	
Karharbari	
Talchir	
Vindhyan	
Sub-metamorphic Series	Proterozoic
Shale, slate, phyllites	
Mica schist, hybrid mica schist	
Quartzite, quartz-schist, quartz mica schist	Archean
Hornblend schist, epidiorite, amphibolite	
Anorthosite	
Granite	
Granite-gneiss	
Porphyritic granite-gneiss	

Source: Singh (1997)

the edge of the Ranchi, Hazaribag, and Koderma Plateaus regions. The Koderma Hills are generally made of Dharwar-age rocks (Mondal et al., 2018). The demarcation of the northern and southern trough boundary in the basin is based on faults that are trending towards east, west and parallel to the Damodar River flows (Tiwary & Dhar, 1994). The present hilly relief feature has been developed by different erosional processes on Gondwana rocks, depending on the hardness and resistance of rocks to erosion (Singh, 1997).

(b) **Plateaus**

Many plateaus have been found in the upper basin area, such as the Ranchi, Hazaribag, and Koderma Plateaus (Fig. 2.2), which are subdivided into the upper, middle, and lower plateaus (Singh, 1997; Mondal et al., 2018). All of these plateaus have more or less similar landform features (DVC, 1992). They are flat with many gullies developed over the plateau area. Isolated low round hills and monadnocks are also found over the plateau region (Mahadevan, 2002). All the sides of the plateau are bordered by steep escarpments in the plateau regions. Monadnocks are a common feature in the Ranchi Plateau. A smaller number of monadnocks are found on the Hazaribag Plateau, while very few are on the Koderma Plateau. Different erosional surfaces with relief features are developed over the plateau regions by the peneplanation process or

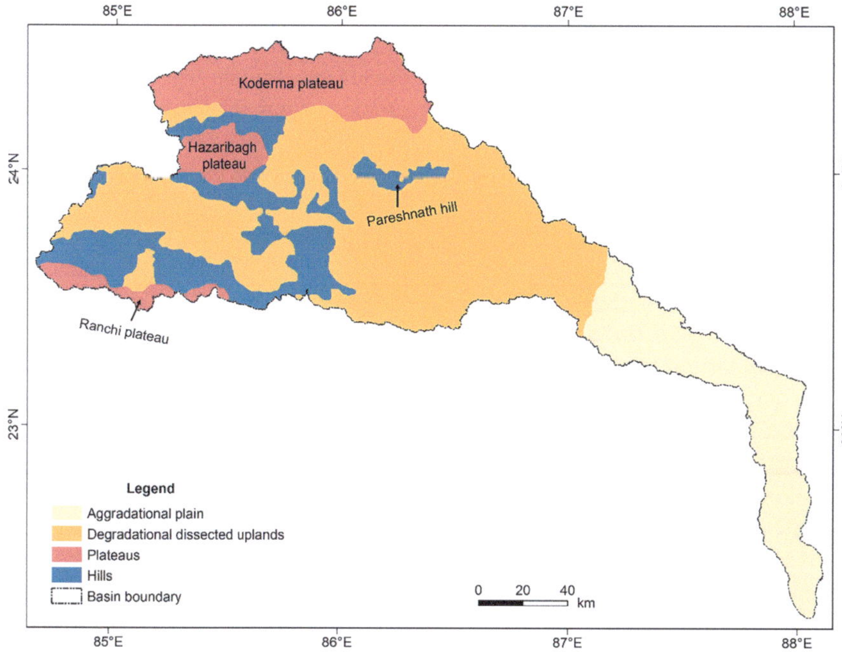

Fig. 2.2 Physiographic divisions of the basin area based on Bhattacharyya (2011)

due to the mineralogical influence of parent rocks (DVC, 1992). The peneplanation process and rock resistance to erosion are almost perfect and uniform in the plateaus region. The Ranchi Plateau has three level erosional features at 15–50 m, 50–100 m and 100–200 m relative relief surface with rounded isolated residual hills (Singh, 1997). The rounded shape was arched due to exfoliation. The relative relief of the Hazaribag and Koderma Plateau is less than that of the Ranchi Plateau. The relative relief of the Hazaribag Plateau is 100–200 m, while the Koderma Plateau is 50–100 m (Mondal et al., 2018). Most of the southern and northern streams originated from these plateau regions (DVC, 1992).

(c) **Degradational and dissected uplands**

The degradational and dissected uplands are found over the upper and middle basins (Fig. 2.2). These areas cover largely Purulia, parts of *Paschim* (West) Bardhaman and Bankura districts in West Bengal and almost the whole Dhanbad and parts of Bokaro, Jamtara, Gridih, Deoghar, Ranchi, Ramgarh, Hazaribagh, Chatra, Latehar districts of Jharkhand. These types of landforms are found along the major river valley from west to east (DVC, 1992), the northern and southern edges of Parasnath Hill, around the Panchet and Maithon reservoirs and the older peneplains (Mondal et al., 2018). These types of landforms are arched from ancient platforms (Singh, 1997) and formed by ancient crystallines, schists, gneisses, and quartzites with basic intrusive rocks in different places (Ghose, 1983). The relative relief of these types of landforms is higher towards the north and west (Mondal et al., 2018). The dominant parts of these degradational uplands lie within the Gondwana trough. The Damodar and Barakar Rivers pass through several tectonic troughs. These troughs are covered by the deposition of Gondwana sediments, and they lie on Archean granite-gneissic basement (DVC, 1992). The eastern edge of the degradational and dissected uplands is filled by large areas of laterites and lateritic soils that lie on the gneissic Archean basement. The laterite soil over the degradational and dissected uplands is the result of fluctuation of the water table and alternate dry and wet seasons. Sheet erosion on the degradational and dissected uplands forms gully and badland topography over the region (Singh, 1997).

(d) **Aggradational flat plains**

Aggradational flat plains are found in the parts of the middle basin and the whole part of the lower basin (Fig. 2.2). These flat plains cover mostly Howrah, Hooghly, and *Purba* (East) Bardhaman and parts of the *Paschim* (West) Bardhaman and Bankura districts of West Bengal. This region is separated from the west degradational and dissected uplands by the line joining Arambagh, Bardhaman, and Bolpur (DVC, 1992). These aggradational flat plains are mostly covered by older and newer alluvial soils in the middle and lower basins. The relative relief of this zone is below 15 m (Mondal et al., 2018). The western part is very much higher than the south-east. It ranges between 15 m in the south-east to 30 m in the west (Singh, 1997). The aggradational plains are to be further subdivided into five zones, such as (a) the Bank plain to the north, (b) the Damodar plain to the south, (c) the Hooghly-Rupnarayan Doab, (d) the Hooghly levee, and (e) the marshes (DVC, 1992).

2.1.1.3 River System

The Damodar River is a subsystem of the Ganga River system of India (Hoque & Islam 2024). The rain-fed Damodar River rises from the Khamarpet Hill of Chotanagpur Plateau at Chulhapani village in Chandwa Community Development (CD) block of Latehar district of Jharkhand at an elevation of 1050 m (Ghosh, 2011), approximately at 23°37′ N and 84°41′ E (Bhattacharyya, 2011; Ghosh, 2014) and meets the Bhagirathi-Hooghly River near Garchumuk, Chandipur, West Bengal (Fig. 2.3). During its journey from Khamarpet Hill to Garchumuk, it changes course abruptly at an almost 90° angle near Chanchai in the *Purba* Bardhaman district of West Bengal. Then near Paikpara, it bifurcates into the Damodar channel (also known as the Amta channel) and the Kanka-Mundeshwari channel. The Damodar or Amta channel joins with the Bhagirathi-Hooghly River, and the other Kanka-Mundeshwari channel joins with the Rupnarayan River. The total length of the river from its origin to the meeting point of the river Bhagirathi-Hooghly is about 592 km (Ghosh et al., 2022; Hoque et al., 2025b). From its source to the confluence point with the Barakar River, the stretch of the Damodar River is considered to be the youthful stage or upper reach. While the stretch from the confluence point of

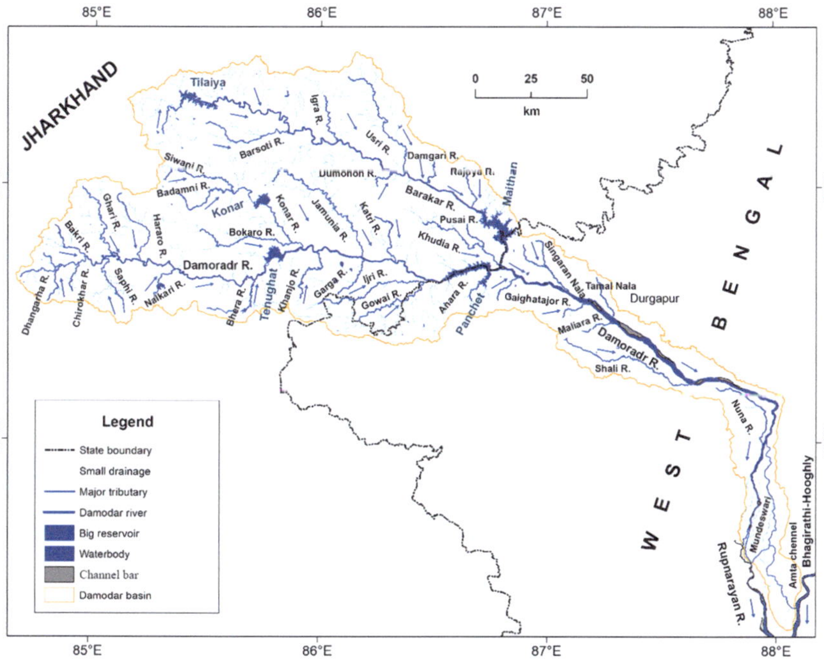

Fig. 2.3 River systems of the Damodar based on SRTM DEM (30 m) and Google Earth Engine

Barakar to Barsul is the middle reach or mature stage, and from Barsul to the confluence point with the Bhagirathi-Hooghly is the lower reach or old stage of the river. The river slope is about 1.86 m/km during the first 241 km, about 0.57 m/km during the next 167 km, and about 0.16 m/km during the last segment (Bhattacharyya & Wiley, 2014).

Originating from the Koderma Plateau region at 640 m elevation, the Barakar River, a principal tributary of the Damodar River, drains through the Hazaribag, Giridih, Kodarma, and Dhanbad districts of Jharkhand and joins near Dishergarh in *Paschim* Bardhaman district (Singh, 1997). The Barsoti (right) and Usri (left) are the main tributaries of the Barakar River. The other medium or small tributaries of the Barakar River are Igra, Dumoho, Damgari, and Rajoya. The Barakar River Basin is a sub-basin and part of the DRB covering a 6293 km^2 catchment area (Sardar et al., 2014).

Besides the Barakar River, Nalkari, Bhera, Saphi, Chirokhar, and Dhangarha are the southern or (right side) tributaries of the Damodar River originating from the Ranchi Plateau, and Bakri, Ghari, and Haharo are the northern or (left side) tributaries originating from the Hazaribagh Plateau (Mondal et al., 2018). These streams flow over the Achean granite-gneissic surface and enter the Gondwana basin (Singh, 1997). The streams Konar and Bokaro originate from the Hazaribagh Plateau and flow through the Archean gneiss surface, but after some distance, Bokaro enters the Gondwana basin at Bokaro coalfields (Mondal et al., 2018). Both streams join at Bokaro thermal station, after that, it (Bokaro) meets with the main river Damodar at Jarandih near Tenughat dam. From Tenughat to Panchet, the river flows eastward. In this section, Jamunia, Katri, and Khudia stream from the north joins the Damodar River while Khanjo, Garga, Ijri, Gowai, and Anara streams from the south join the Damodar River (DVC, 1992). Barakar and Bokaro are two major left bank tributaries of Damodar in Jharkhand, and Shali is the major right-bank tributary of the river in Bankura district of West Bengal. Other primary and secondary rivers of the Damodar are given in detail in Table 2.3. Along with Harinkhola, short-cut channel, *Kana* Dwarakeswar, and Hurhura *Khal* are other important drainage veins of this catchment, playing a vital role in draining out flood discharge into the river Rupnarayan, which has tidal influence.

2.1.1.4 Climate

According to Stamp's climatic classification of India, the Damodar River Basin lies in a heavy rainfall region. The mean annual temperature ranges from 18° to 24°C in winter and 29–35°C in summer. The annual average rainfall varies from 100 to 200 cm. According to Koeppen's scheme of climatic classification of India, the upper basin, mainly the Chotanagpure Plateau region, lies in the monsoon type with dry winters (Cwg) and 100–200 cm rainfall. The lower Damodar lies in the Tropical

Table 2.3 Details of the Damodar River system

Tributaries		Location	
Primary	Secondary	State	District
Barakar	-	Jharkhand	Hazaribag, Giridih, Kodarma, Dhanbad
	Igra		Giridih
	Ushri		Giridih
	Dumohon		Giridih
	Barsoti		Hazaribag
Barki	–		Latehar, Chatra, Hazaribag
Haharo	–		Hazaribag
Ghari	–		Hazaribag
Bokaro	–		Hazaribag, Bokaro
Konar	–		Hazaribag, Bokaro
	Siwani		Hazaribag
Jamunia			Hazaribag, Giridih, Bokaro, Dhanbad
Naikari, Bhera			Ranchi, Ramgarh
Khanjo, Garga			Bokaro
Khadia, Katri			Dhanbad
Gowai, Ijri			Bokaro
		West Bengal	Purulia
Sali			Bankura
Singar, Tamal			Bardhaman
Nuna			

Source: Annual flood report (2017), Govt. of West Bengal

Savanah type (Aw) with 75 cm of rainfall. The climate of the basin is influenced by the Indian tropical monsoon. Tropical wet and dry with dry winter seasons prevail in the entire basin. The climate of the basin is characterized by tropical dry winters with hot and humid summers (Chandra, 2003). In the winter season, the air becomes dry, and very little rainfall occurs from December to March, influenced by western disturbances. In the winter, the north east winter monsoon blows with less humidity. In the summer, the wind blows from sea to land in the south-west direction with high humidity clouds and rain, called the south-west monsoon. The air becomes humid and rainfall occurs, influenced by the south-west monsoon from June to September. Apart from that, the transition seasons are also observed, such as the hot months of April and May and monsoonal retreating in October and November. Early rainfall also occurs in April and May due to conventional thunderstorms. In West Bengal, it is called *'kalboishaki'*. The Damodar and its drainage system are rain-fed by the rainfall which mainly occurs in the period of monsoon, June to September and drains out as surface runoff in the Damodar basin area.

(a) **Pattern of temperature**

Generally, the high temperature prevails between May and June when the maximum temperature reaches 42°C in summer, while December and January exhibit the cold months when the temperature strikes to 26°C. The highest maximum temperature crossed 46°C as recorded over a greater part of the basin. In May average maximum temperature is 43°C and the average minimum temperature is 30°C. December and January are the coldest months in the winter season and temperatures in these months fall below 4°C at some locations in the basin. Mean relative humidity in the air differs from 40% during March, April and May to 80% from July to September (Chandra, 2003). The spatial variation of mean temperature in DRB during 2011–2020 is represented in Fig. 2.4.

(b) **Distribution of rainfall**

The average annual rainfall of the basin ranges from 1200 to 1600 mm. Moreover, the average annual rainfall was 1343 mm from 1901 to 2002 in the basin area (Table 2.4).

The distribution of rainfall in the basin extensively varies based on the topographic and climatic conditions in the different parts of the basin. The average annual rainfall is 1282 mm in the upper basin, 1390 mm in the middle basin, and 1570 mm in the lower basin. The average annual rainfall of the basin is 1335 mm, and 82% of rainfall occurs during the monsoon period (June to September) (Ghosh et al., 2016). The distribution of annual rainfall in DRB is represented in Fig. 2.5 during 2011–2020.

2.1.1.5 Soil System in the Damodar River Basin

The soil of the basin has been grouped into three major types such as red loam, laterites, and lateritic and riverine alluvial (Fig. 2.6). The red loam type of soil is found in the upper and middle reaches of the basin. The laterites and lateritic soil are

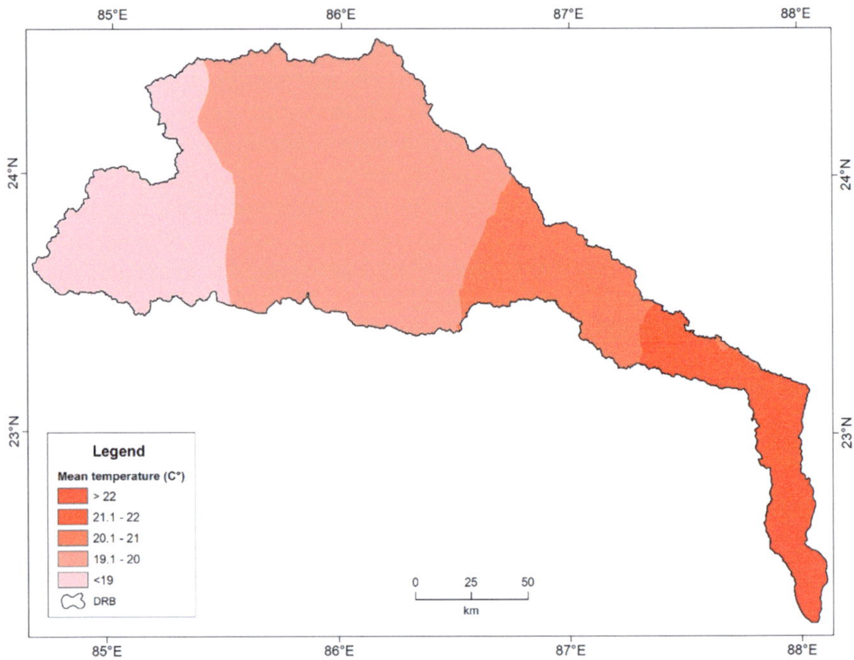

Fig. 2.4 Distribution of temperature in the Damodar River Basin during 2011–2020 based on Climatic Research Unit gridded Time Series (2023)

Table 2.4 District-wise rainfall distribution in the Damodar River Basin (1901–2002)

Sl. No.	Sub-region	Districts	Average rainfall (mm)
1	Jharkhand	Bokaro	1357.84
2		Jamtara	1345.93
3		Ramgarah	1305.13
4		Lohardaga	1294.56
5		Latehar	1224.72
6		Chattra	1211.05
7		Deoghar	1312.41
8		Kodarma	1255.67
9		Ranchi	1299.50
10		Hazaribagh	1305.13
11		Giridih	1300.32
12		Dhanbad	1340.93
1	West Bengal	Purulia	1318.25
2		Bankura	1411.34
3 and 4		Both Bardhaman	1417.09
5		Hooghly	1540.86
6		Howrah	1602.46

Source: Based on India Water Portal, Tyndall Centre for Climate Change Research (2022)

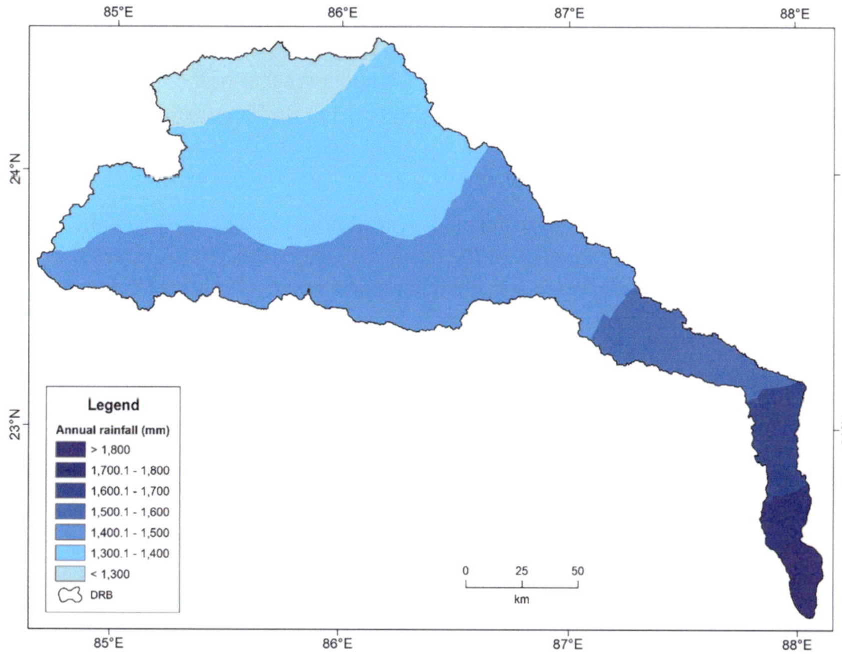

Fig. 2.5 Distribution of rainfall in the Damodar River Basin during 2011–2020 based on Climatic Research Unit Gridded Time Series (2023)

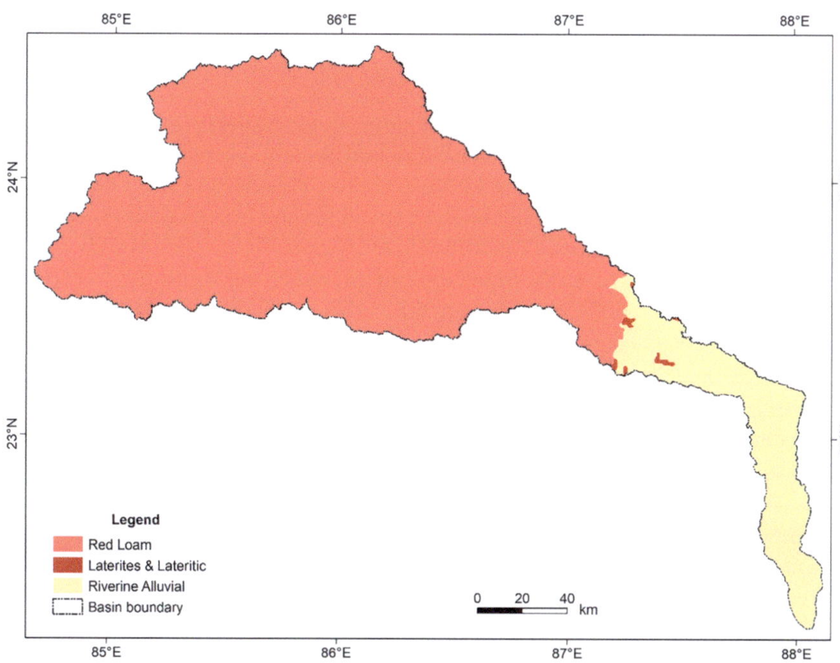

Fig. 2.6 Soil in the Damodar River Basin based on Bhattacharyya (2011)

found in the middle reach of the basin, developed on the degradational and dissected uplands (Singh, 1997). Riverine alluvial soils are mainly found in the middle and lower reaches and also in a small section of the upper reaches of the basin. The basin soils are to be classified into the following types: (1) Red loam soils, (2) Laterites and lateritic soils, and (3) Riverine alluvial soils based on Bhattacharyya (2011).

(a) **Red loam soils**

Red loam soils have been formed over almost the whole part of the upper reach and upper middle reach of the basin, where granite, gneiss, or schist is the predominant or parent rock and the denudation process is active (Singh, 1997). These soils are found in all the districts of the Jharkhand sub-region (Table 2.4) and parts of *Paschim* Bardhaman and Purulia districts of West Bengal. High temperatures with moderate humidity are favourable conditions for developing these types of soils. The active denudation process with favourable climatic conditions and primary rock is the composite factor of pedogenesis in the region. These types of soils appear reddish in colour, resulting in oxidation in ferrous materials under climatic and vegetative environment conditions of the region (DVC, 1992). The soils are formed by the laterization process, the organic materials of the upper layer, or 'A horizon' are leached out and deposited in the lower layer, or 'B horizon' and only ferrous materials remain in the upper layer (illuvium). The alternating dry and wet climate initiates the lateralization of these soils. The undulating topography and deforestation lead to accelerating soil erosion in the region. These types of soils have very little organic material, nitrogen, and phosphorus but are highly concentrated with potash content (DVC, 1992). This type of soil's texture varies from loamy sand to sandy loam. These types of soils have a thin layer and they are less fertile (Singh, 1997).

(b) **Laterites and lateritic soils**

Laterites and lateritic soils are developed in the degradational and dissected plateau fringes in Bardhaman and Purulia districts in the west. These types of soils are formed by weathering under oxidation and leaching processes. These types of soils are formed under high temperatures and heavy rainfall with alternating wet and dry climatic conditions in the region. These types of soils appear yellow-grey in colour on the top of the soil and red colour below the soil (DVC, 1992). These types of soils have very little in organic matter, nitrogen, available phosphorus, and calcium content but are rich in iron and aluminium (Mondal et al., 2018). These types of soils are less fertile due to a lack of organic matter, nitrogen, phosphorus, and potassium. They are porous and claylike (Olutaiwo & Olushola, 2017).

(c) **Riverine alluvial soils**

Riverine alluvial soils are found in the whole lower reach and lower middle reach of the basin. The alluvial soil is formed by the deposition of the erosional material or sediments brought by the river system and deposited on the lower middle reach and tail reach of the basin. There are two types of alluvial soils found based on depositional characteristics, such as older alluvial soils and newer alluvial soils (DVC, 1992). The old alluvium soils are found in the lower middle reach and levee region and on diverse alluvial terraces. The older alluvial soils are observed over the levee region and on the alluvial terraces at different heights (Mondal et al., 2018). These types of soil are generally red in colour, being iron-rich. The newer alluvial soils are extensively found in the floodplain region of the lower reach of the basin. The new alluvial soils cover a wide area in Birbhum, Bankura, and Bardhaman districts of West Bengal. The texture of the soil is moderate to very coarse. In the low-lying tracts of Hooghly and Howrah districts, marshy or swampy soils are also found (DVC, 1992). These types of soils are rich in minerals, nutrients, and organic matter and also hold heavy metals and are highly fertile.

2.1.1.6 Nature of Vegetation

Several types of vegetation and their species are found in this basin. It covers open and dense forests and open grasslands. It covers less than 20% of the basin area, below the national average (DVC, 1992). It has the spatial extents of vegetation cover in the upper, middle, and lower reaches of this basin. The lower reach of this basin is covered by less than 3% area by vegetation; however, 40% of the area of the upper basin is covered by vegetation. A large strip of the upper and middle reaches of this basin has been deforested due to mining, quarries, and industrial setup (Singh, 1997). The upper reach is a good source of wood and preserves woodlands. Vegetation of this basin is varied with climate, edaphic, and topographic conditions of the land. Deciduous and semi-deciduous vegetation is prevalent in this basin. The tropical moist deciduous vegetation is found in the area where the rainfall surpasses 1250 mm and the dry deciduous vegetation types are found in drier parts where the rainfall is below 1250 mm in the basin (DVC, 1992).

This basin is rich in floral biodiversity with 137 flowering plant families and 853 species belonging to 535 genera (Choudhury et al., 2012). The *Poaceae,* with its 148 species, is the leading family and *Leguminoseae* with its 92 species, is the second largest family in this basin. Various types of medicinal plants found in this basin (Banerjee, 2013) provide benefits to the local population. Sal *(Shorea robusta)* is the dominant plant among all the available species. It dominates in the areas characterized by moist and low-lying conditions with relatively deeper soils. It is associated with Asan *(Terminalia tomentosa),* Kend *(Diospyros melanoxylon),* Garnahr *(Gmelina arborea),* etc. in the protected valleys and basins with deep soil covers. But in a drier climate, trees dominate over Sal in the region of steeper slopes with thin soil cover. Other species, such as Salai, Dhak, Aonla *(Phyllanthus emblica),* and bamboo, are common in this basin. Accompanying the Sal and other species like Mahua *(Bassua latifolia)* and Jarnun *(Eugenia jamobalana),* mixed forests are common in the region. Several species of grass, such as Sabai and Spear, are also found in the moist deciduous forest in this basin (DVC, 1992). Open/dispersed and dense Sal forest (*Shorea robusta*) is found on laterite, red, and yellow loamy soil in the upper and middle reaches of this river basin.

2.1.2 Cultural Aspects

2.1.2.1 Population and Settlements

The population of the basin was 14.66 million reported in the census of India (2001), while it was 17.31 million as per the census (2011). The population growth rate of this basin is 16.15% from 2001 to 2011. The basin area is covered by 12 districts of Jharkhand state and six districts of West Bengal state. The district-wise population and their growth in the basin area are given in Table 2.5. The population density in the basin area was 692.49 persons per km² in 2001 and 806.58 persons per km² in 2011 (Table 2.6). The distribution of population density in the basin area is given in Table 2.6.

2.1.2.2 Transport and Communication

The Damodar River Basin region has both types of transport systems, i.e., road and rail (Fig. 2.7). Here, the main railway system is the Asansol-Gaya railway section, Dhanbad-Adra, Asansol-Adra, Howrah-Jabalpur, Asansol-Tatanagar-Kharagpur line, Howrah-Kharagpur line, Bardhaman-Asansol railway section, Howrah-Goghat line, Howrah-Bankura line and Grand Chord railway path.

The important roads are Grand Trunk (GT) Road, NH 16, NH-18, NH-19, NH-20 (Ranchi-Patna Road), NH-22, NH-39, NH 114A, NH 116A, NH-218, NH-320, NH-419, and NH-522. These roads and railways are important for communicating with the metropolitan cities of Kolkata, Delhi and other cities in India. Therefore, this type of good transportation system helps connect with other industrial regions and other big towns in the country, which is helpful in the development of this industrial region.

2.1.2.3 Land Use and Land Cover

In the study area, eight major types of land use and land cover (LULC) features are found. Agricultural land, vegetation (tree), and built-up area are the dominant LULC, covering 86.78% of the basin area. Agricultural land has the highest share, covering about 50% of the basin area. Vegetation cover (22.73%) is mainly found in hilly regions in the upper part of the basin. The built-up area covers 13.18% of the basin area, and the densely built-up area is mostly observed along the riversides. The mining area is also observed along the river sides in the upper part of the basin (Fig. 2.8).

Waterbody covers 2.65% of the basin area. The waterbody area is mainly shared by the dam-reservoir area in the upper part of the basin. The distribution and details of LULC in the basin are shown in Fig. 2.8 and Table 2.7.

Table 2.5 District-wise population details in the basin area

Sl. No.	Districts	Total geographical area of the district (km²)	District area within the basin (km²)	District area in basin (%)	Basin area share in district (%)	Population in the basin area		
						(2001)	(2011)	Growth (%)
Jharkhand Sub-Region								
1	Bokaro	2868.38	2842.56	99.10	12.16	1,761,660.23	2,043,765.76	16.01
2	Jamtar	1803.34	539.13	29.90	2.31	195,246.53	236,491.56	21.12
3	Ramgarh	1343.69	1200.70	89.36	5.14	750,148.51	848,408.01	13.10
4	Lohardanga	1500.77	58.84	3.92	0.25	14,292.78	18,106.67	26.68
5	Latehar	4268.34	692.87	16.23	2.96	91,048.48	118,008.47	29.61
6	Chatra	3758.52	645.90	17.19	2.76	136,008.36	179,220.52	31.77
7	Deoghar	2439.08	6.68	0.27	0.03	3190.37	4084.70	28.03
8	Kodarma	1321.93	546.42	41.33	2.34	206,426.87	296,063.71	43.42
9	Ranchi	4883.44	806.83	16.52	3.45	388,364.13	481,487.94	23.98
10	Hazaribagh	4816.40	4164.10	86.46	17.82	1,242,924.15	1,499,587.32	20.65
11	Giridih	4993.67	3492.45	69.94	14.94	1,331,910.86	1,710,303.55	28.41
12	Dhanbad	2082.31	2082.31	100.00	8.91	2,397,102	2,684,487	11.99
Sub-Total	–	**36,079.87**	**17,078.80**	**47.34**	**73.08**	**8,518,323.27**	**10,120,015.31**	**18.80**
West Bengal Sub-Region								
1	Purulia	6252.38	1398.09	22.36	5.98	567,188.89	655,201.33	15.52
2	Bankura	6959.33	1557.59	22.38	6.66	714,569.01	804,985.06	12.65
3 and 4	Both Bardhaman	7037.15	2102.17	29.87	8.99	2,059,857.09	2,305,423.04	11.92
5	Hooghly	3150.58	490.65	15.57	2.10	785,201.36	859,512.25	9.46
6	Howrah	1439.65	743.34	51.63	3.18	2,206,344.04	2,504,232.32	13.50
Sub-Total	–	**24,839.09**	**6291.84**	**25.33**	**26.92**	**6,333,160.39**	**7,129,354.00**	**12.57**
Total	–	–	**23,370.64**	–	**100**	**14,851,483.66**	**17,249,369.31**	**16.15**

Source: Based on the census (2001 and 2011)

Table 2.6 District-wise population density in the basin area

Sl. No.	District	Population density (persons per km²)	
		2001	2011
Jharkhand sub-region			
1	Bokaro	619.74	718.99
2	Jamtar	362.15	438.66
3	Ramgarh	624.76	706.59
4	Lohardanga	242.89	307.70
5	Latehar	131.41	170.32
6	Chatra	210.57	277.47
7	Deoghar	477.80	611.74
8	Kodarma	377.78	541.83
9	Ranchi	481.34	596.76
10	Hazaribagh	298.49	360.12
11	Giridih	381.37	489.71
12	Dhanbad	1151.17	1289.19
West Bengal sub-region			
1	Purulia	405.69	468.64
2	Bankura	458.76	516.81
3 & 4	*Purba* and *Paschim* Bardhaman	979.87	1096.69
5	Hooghly	1600.33	1751.79
6	Howrah	2968.15	3368.89

Source: Based on the census (2001 and 2011)

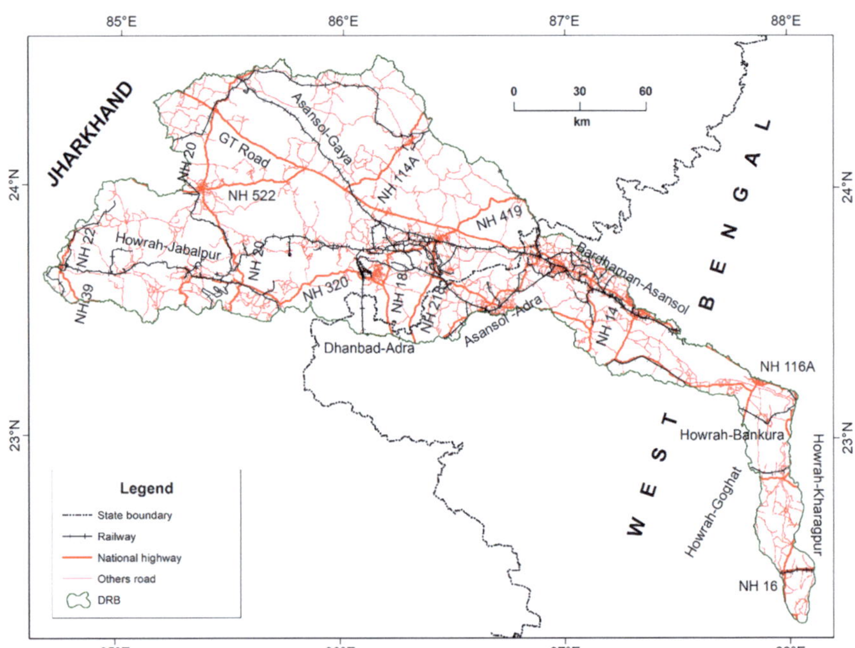

Fig. 2.7 Transport and communication system in the Damodar River Basin

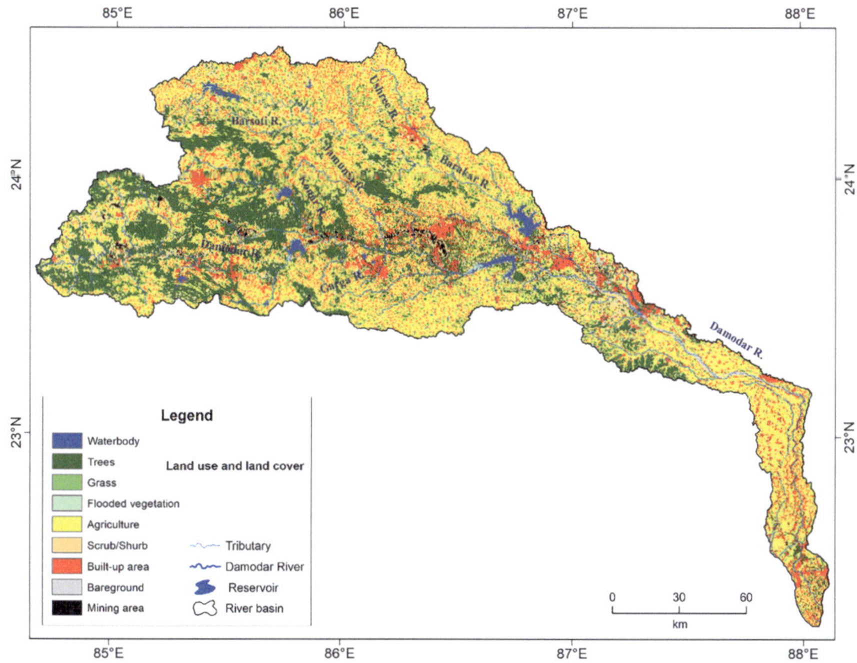

Fig. 2.8 Land use and land cover map of the Damodar River Basin

Table 2.7 Details of land use and land cover in the Damodar River Basin

Sl. No.	Classification	Area (km²)	Area (%)
1	Waterbody	620.49	2.65
2	Tree	5320.65	22.73
3	Grass	72.49	0.31
4	Flooded vegetation	36.09	0.15
5	Agriculture	11,908.60	50.87
6	Scrub/Shrub	2286.67	9.77
7	Built-up area	3086.62	13.18
8	Bareground	80.33	0.34

Computed by the authors based on the LULC map of 2020

2.2 Floods in the Damodar River Basin

Flood is a natural phenomenon, and it leads to hazards. It is the most devastating, widespread, and frequent among all natural hazards (Ghosh et al., 2022; Sanyal & Lu, 2004). It results from extreme rainfall within a short period or time in the catchment area. Sometimes, a sudden flood also occurs, called a flash flood. This type of flood with dangerous head waves is also known as '*hurpaban*' (Bhattacharyya, 2011). The rain-fed Damodar River and its basin are flood-prone, influenced by the south-west monsoon (Hoque et al., 2023). Heavy rainfall and about 80% of the river's annual discharge occur in monsoon months, i.e., June to September. The heavy monsoonal rainfall with high discharge water leads to floods in the basin area. The maximum flood occurs from August to September as per the flood history of DRB. The flood consequences are public property and life loss, and sometimes it changes the river course. The devastating consequences of the Damodar River floods were delivered into folklore, for example, the Bhadu song as-

We have sown the crops in Asar
We will bring Bhadu in Bhadra.
Floods have swollen the Damodar
The sailing boats cannot sail.
O Damodar! We fall at your feet
Reduce the floods a little.
Bhadu will come a year later.
Let the boats sail on your surface.

A flood is a hydro-geomorphological process in a physical environment, and it is controlled by hydro-meteorological factors (mainly rainfall) of the natural environment (Bhattacharyya, 2011). The Damodar River is called '*The Sorrow of Bengal*', the term derived from its characteristics, as several widespread floods occurred in the basin area. This river has a long flood history as per the recorded track from 1730 to the present (Bhattacharyya, 2011), although it has been traced from 1665 to 2020. The flood history spans 355 years (1665 to 2020). The first recorded flood in the basin was in 1730, but an unspecified poet had also mentioned the flood of 1665 (Bhattacharyya, 2011). Since 1730, every 8–10 years, several floods have occurred with a variety of intensity and magnitude (Bhattacharyya, 2011). Sometimes, sudden floods would occur with hazardous head waves called '*hurpaban*' (Hunter, 1876; Bhattacharyya, 2011). This type of flood would have a great breadth and look like a wall of water, sometimes 1.5 m in height, resulting in severe destruction of land, properties, and people (Hunter, 1876). Most serious and devastating floods occurred in the year 1730, 1770, 1823, 1840, 1848, 1855–56, 1866, 1873–74, 1875–76, 1882, 1884–85, 1891–92, 1897–98, 1900–01, 1905, 1907, 1913, 1916, 1923, 1927, 1930, 1935, 1940–41, 1942–43, 1946, 1950–51, 1956, 1958–59, 1978, 1984, 1991, 1995, 1999, 2000, 2006, 2007, 2009, 2015 and 2020 in the basin area. Moreover, flood frequency analysis represents that on average 5 devastating floods occurred in 25 25-year interval and the highest, 9 devastating floods occurred between 1876–1900 and 1926–1950 as per the track record from 1801–2022 (Fig. 2.9).

Furthermore, as per the flood records track, 37 times floods occurred with a peak flow of ≥8496 cusecs in the years 1823 to 2007 (Bhattacharyya, 2011). Floods occurred with >8496 cumecs peak flow in the year 1823. 1840, 1855, 1860, 1864. 1865, 1866, 1877, 1913, 1916, 1917, 1935, 1938, 1940, 1941, 1942, 1946, 1950, 1951, 1956, 1958, 1959, 1978, and 2007 (Bhattacharyya, 2011; Ghosh, 2014). However, floods occurred with >16,992 cumec peak flow in the years 1823, 1840, 1913, 1935, 1941, 1958, 1959, and 1978. Floods with about 18,678 cumecs peak flow were recorded in August 1913 and 1935, and October 1941 (Bhattacharyya, 2011). The most devastating floods of 1913, 1935, and 1943 occurred with high peak flow (18,678 cumecs) because of heavy rainfall in the basin area. The 1913

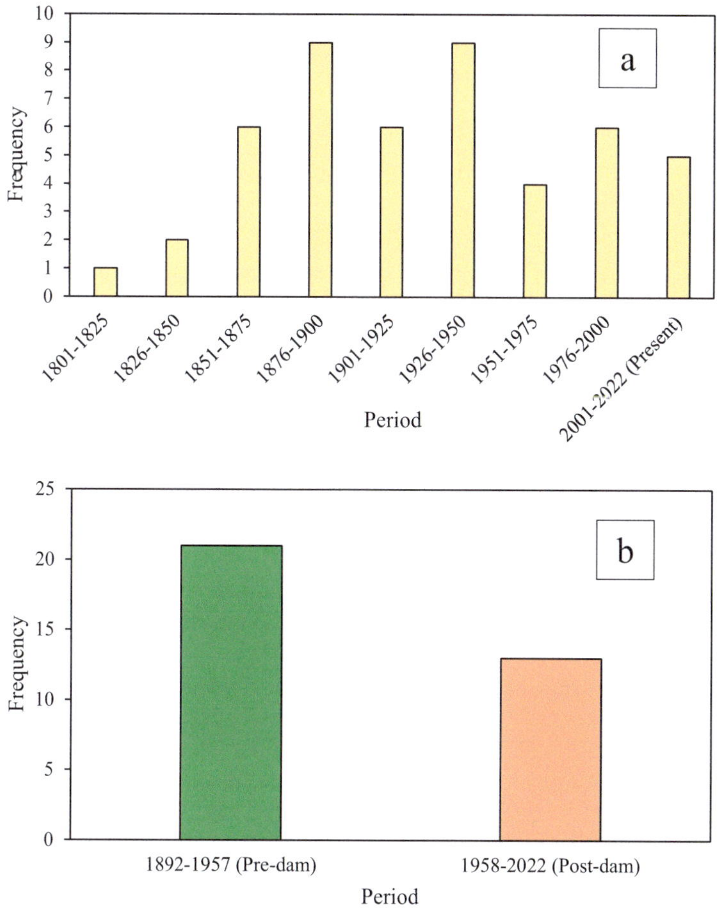

Fig. 2.9 Flood frequency analysis from 1801–2022; (**a**) 25 years interval and (**b**) during 65 years in pre-and post-dam period

flood occurred due to 30.23 cm of mean rainfall in a day over the basin area. The 1935 flood occurred for 23.29 cm of mean rainfall in 3 days. While the flood of 1941 occurred for 21.34 cm mean rainfall with relatively low peak flow (9911 cumecs). Although low peak flow, the destruction was highest as recorded based on the prices prevailing in 1951 (Bhattacharyya, 2011). These endemic floods led to disaster as the countryside was inundated, crops and domestic animals were washed out, lots of life lost, and transportation and communication systems were badly impacted. For example, the floods between 1862 and 1872 resulted from a malaria epidemic and about one-third of the population died on the left bank of the river, known as 'Bardhhaman fever'. The flood of 1943 was more devastating because it submerged the countryside of south Bardhhaman and inundated 6–7 ft in some places, swept away villages, destroyed rice fields, disrupted rail traffic, washed away the railway bridge, breached the grand trunk road and also created a dangerous situation for

Bardhaman and Kolkata cities for a long time. The magnitude of damage was higher in some years and lower in others. Most of Bardhaman town was flooded in the years 1770, 1855, 1913, and 1943.

2.2.1 Pre-dam Flood

Before the installation of the dam, the most serious and devastating floods occurred in the year 1730, 1770, 1823, 1840, 1848, 1855–56, 1866, 1873–74, 1875–76, 1882, 1884–85, 1891–92, 1897–98, 1900–01, 1905, 1907, 1913, 1916, 1923, 1927, 1930, 1935, 1940–41, 1942–43, 1946, 1950–51, and 1956 in the basin area. From 1823, floods occurred with >8496 cumecs peak flow in the year 1823. 1840, 1855, 1860, 1864. 1865, 1866, 1877, 1913, 1916, 1917, 1935, 1938, 1940, 1941, 1942, 1946, 1950, 1951, and 1956 (Bhattacharyya, 2011; Ghosh, 2014). Whereas, they occurred with >16,992 cumec peak flow in the years 1823, 1840, 1913, 1935 and 1941. About 18,678 cumecs peak floods had been recorded in August 1913 and 1935 and October 1941 (Bhattacharyya, 2011). Moreover, before 65 years (1892–1957) of the dam periods, 21 devastating floods happened in the basin according to the flood record (Fig. 2.9). Besides, the flood frequency study at Rhondia during 23 years shows that 2 extremely abnormal (>12,744 cusec), 7 abnormal (>8496 cusec), 11 normal (5664–8496 cusec), and 4 subnormal (<5664 cusecs) floods occurred in 1933–1956 (Fig. 2.10 and Table 2.8). The flood occurrence and magnitude or intensity in the pre-dam period were high compared to the post-dam periods. Furthermore, flood frequency analysis from the Log Pearson probability distribution shows that the return period of 1.05 years with the corresponding 95.2% flood probability discharge is 3595 m^3/s in the pre-dam condition (1823–1957). In this period, 5 years return period has the flood probability peak discharge of 12,922 m^3/s. Moreover, 100 and 200 years return periods of flood probability peak discharge are 20,077 m^3/s and 21,201 m^3/s, respectively, in the pre-dam condition (Table 2.9).

2.2.2 Post-dam Flood

After the installation of the dams, the frequency and magnitude or intensity of floods have been reduced substantially in the basin areas. However, some serious and destructive floods also happened in the basin in the years 1958–59, 1978, 1984, 1991, 1995, 1999, 2000, 2006, 2007, 2009, 2015, and 2020 in the post-dam period. Floods with >8496 cumec peak flow occurred in 1958, 1959, 1978, and 2007, while they occurred with >16,992 cumec peak flow in the years 1958, 1959, and 1978 (Bhattacharyya, 2011; Ghosh, 2014). Besides, 13 devastating floods happened in the basin after 65 years (1958–2022) of the dams (Fig. 2.9). Nearly 40% of these devastating flood occurrences were reduced in the post-dam period as compared to before and after 65 years of dam study. Additionally, the flood frequency study during 50 years at Rhondia reveals that 3 abnormal (>8496 cusec), 5 normal (5664–8496

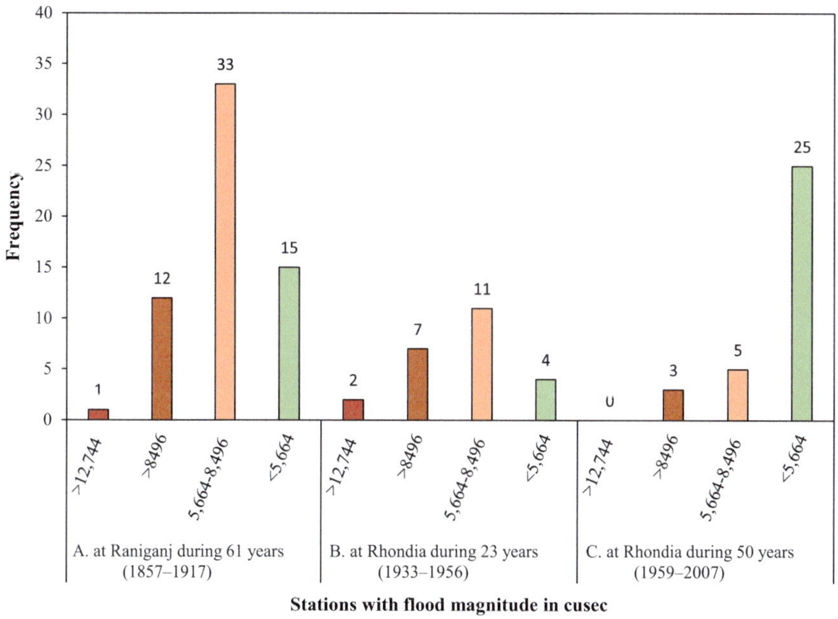

Fig. 2.10 Flood history in the selected stretches of the lower Damodar River

Table 2.8 Flood history in the selected stretches of the lower Damodar River

Stations	Flood intensity	Flood magnitude in cusec	Frequency
A. at Raniganj in 61 years of the pre-dam period (1857–1917)	Extremely abnormal	>12,744	1
	Abnormal	>8496	12
	Normal	5664–8496	33
	Subnormal	<5664	15
B. at Rhondia in 23 years of the pre-dam period (1933–1956)	Extremely abnormal	>12,744	2
	Abnormal	>8496	7
	Normal	5664–8496	11
	Subnormal	2472–5664	4
C. at Rhondia in 50 years of the post-dam period (1959–2007)	Extremely abnormal	>12,744	0
	Abnormal	>8496	3
	Normal	5664–8496	5
	Subnormal	2472–5664	25

Source: Bhattacharyya (2011)

cusec), and 25 subnormal (<5664 cusecs) floods occurred in 1959–2007 (Fig. 2.10 and Table 2.8). These results also indicate that the high peak floods have been reduced substantially, while subnormal floods increase in the post-dam periods at Rhondia. Furthermore, the Log Pearson probability analysis of flood frequency reveals that the 1.05 year return period with the corresponding 95.2% flood

Table 2.9 Log Pearson (1823–1957)

i	Return period T (years)	Probability P (percent)	Frequency factor K	$y = \log(Q)$	Flood discharge Q (m^3/s)
1	1.05	95.2	−1.826	3.556	3595
2	1.11	90.1	−1.333	3.658	4546
3	1.25	80	−0.786	3.771	5901
4	2	50	0.121	3.959	9100
5	5	20	0.856	4.111	12,922
6	10	10	1.177	4.178	15,057
7	25	4	1.474	4.239	17,346
8	50	2	1.642	4.274	18,802
9	100	1	1.78	4.303	20,077
10	200	0.5	1.894	4.326	21,201

Table 2.10 Log Pearson (1958–2007)

i	Return period T (years)	Probability P (percent)	Frequency factor K	$y = \log(Q)$	Flood discharge Q (m^3/s)
1	1.05	95.2	−1.834	2.93	851
2	1.11	90.1	−1.335	3.077	1194
3	1.25	80	−0.782	3.24	1739
4	2	50	0.128	3.509	3229
5	5	20	0.856	3.724	5296
6	10	10	1.17	3.817	6556
7	25	4	1.458	3.902	7971
8	50	2	1.62	3.949	8900
9	100	1	1.751	3.988	9729
10	200	0.5	1.859	4.02	10,469

probability discharge is 3595 m^3/s in the post-dam condition (1958–2007). Again, the 5 years return period has a flood probability discharge of 5296 m^3/s. Additionally, the flood probability peak discharge of 100 and 200 years return periods is 9729 m^3/s and 10,469 m^3/s, respectively in the pre-dam condition (Table 2.10).

2.2.3 *Flood Geomorphological Issues*

2.2.3.1 Morphological and Sedimentological Aspects

Morphological and sedimentological processes with human influence are reflected in changes to channel planform, shape, and roughness, as well as in flow-stage relationships, sediment supply, erosion, and floodplain sedimentation (Fig. 2.11) (Church, 2015; Patel & Kumar, 2017; Vázquez-Tarrío et al., 2024). These processes significantly affect floods by altering river characteristics and raising flood risk. Sediment deposition or increased sedimentation reduces the river's depth, raising

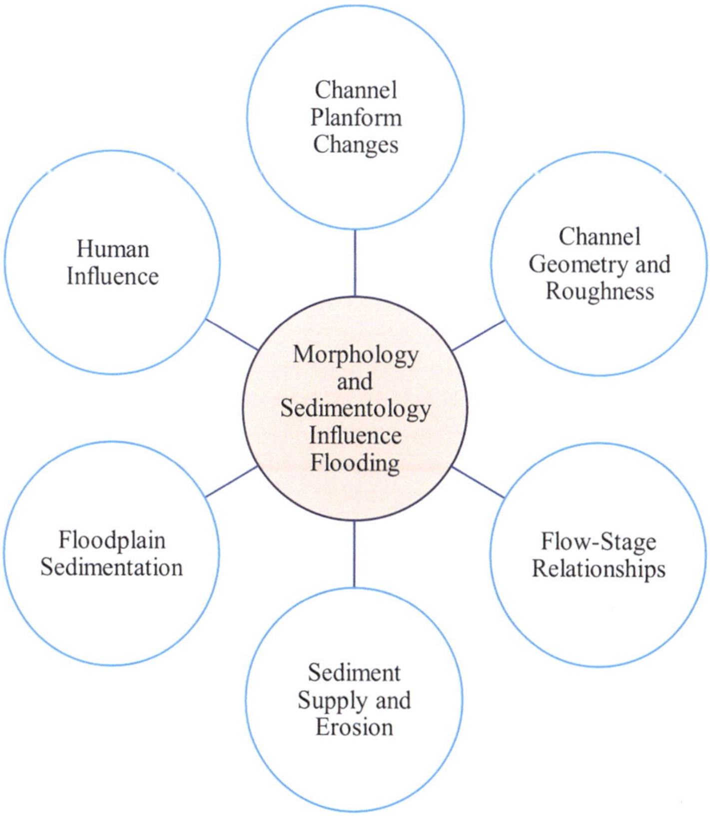

Fig. 2.11 Influence of morphological and sedimentological processes on flooding

water levels and intensifying floods. Sediment deposited on the riverbed changes its cross-sectional shape, decreasing the ability and raising water levels during flood events. Additionally, sediment accumulation on the floodplain during flooding records the event's scale and extent. It also gathers on delta fronts, delays nutrient delivery, and creates new features that influence future flood patterns. Therefore, sediment deposition or sedimentation modifies the river's form and capacity, impacting its ability to hold water, flood dynamics, and risk levels. Conversely, sediment erosion can heighten flood severity by forming new channels or deepening existing ones, affecting the scale and timing of floods.

Moreover, sediment transport causes rapid changes in the river's shape, such as bank erosion leading to channel widening or course alterations through avulsion. These changes affect the river's overall ability to contain floodwaters and influence the relationship between water level and flow rate. Additionally, intense rainfall in catchment areas introduces substantial sediment loads into river systems, resulting in debris flows and flash floods. The erosive power of floods selectively removes fine sediments, impacting the river's capacity to transport and deposit coarser materials, and causing local modifications to riverbeds and floodplains. Human influences, like changes in LULC, further modify runoff, erosion, and sediment supply, making floods more 'flashy' and severe.

2.2.3.2 Lithological Aspects

Lithological characteristics affect water flow by regulating infiltration and flow capacity in an area. Pervious rocks and soils permit more water to percolate or infiltrate, decreasing surface runoff and flood risk. However, impervious rocks and compressed soils hinder infiltration, rising runoff and flood severity. Certain types of rock, such as highly porous sandstone versus impervious rock, determine how much water enters the soil compared to flowing on the surface, directly influencing the volume of water that overcomes drainage systems and leads to flooding. Permeable lithologies (sand, gravel, porous sandstone) enable rainwater to seep into the ground, replenishing groundwater and lowering surface flow. Conversely, impervious lithologies (dense clay, shale) block water infiltration, resulting in higher surface runoff. Lithology shapes the drainage patterns of an area, with some rock formations facilitating more efficient underground drainage, while others promote surface water accumulation. It also directly influences soil type and properties, which impact water absorption. Soils resulting from previous rocks tend to be more porous, whereas those from impervious rocks are less so. Hence, the lithology has a significant impact on flood severity, as impervious lithology-dominated areas are more vulnerable to flash floods due to water cannot percolate into the soil. Whereas areas with porous lithology infiltrate rainfall into the soil, which decreases the risk of flash floods and supports the hydrological response. Therefore, rock properties, soil characteristics and water retention or absorption capacity play crucial roles in determining the flood risk of an area or region. This characteristic dictates how much water stays in the ground and how much becomes surface runoff, which influences the flood probability and severity.

2.2.3.3 Role of LULC in Floods

Land use and land cover (LULC) significantly influence floods by changing the hydrological cycle. It's changes such as deforestation, urbanization, agricultural expansion, mining, and industrial activities, reduce forest cover and infiltration, leading to increased surface runoff, faster peak flows, and higher flood magnitudes and frequency (Khodaei et al., 2025; Abdulkareem et al., 2018). However, intact forest cover, afforestation, and wetlands can absorb rainfall, decrease runoff, and thus lessen the impact of flooding. Conversely, mining activities can increase infiltration and help recharge groundwater by storing water in mining areas, thereby reducing surface runoff, rapid peak flows, and the magnitude and frequency of severe floods. Therefore, LULC changes are a crucial factor in elevating flood risk which highlighting the need for improved land management to mitigate floods.

2.3 Droughts in the Damodar River Basin

Drought, like a flood, is a common phenomenon in India, occurring each year in various parts of the country. The Indian climate is influenced by the south-west and north-east monsoons, with maximum rainfall typically occurring during the monsoon period. However, due to climate change and global warming, rainfall has become unpredictable in India. The DRB has historically faced both floods and droughts because of its unique topography, dependence on the monsoon, and land use patterns. The root causes of drought in the DRB include irregular monsoon patterns, where delays or shortages in rainfall can lead to severe water stress, along with deforestation and land degradation resulting from large-scale mining and industrial activities, particularly in the upper basin. This has led to deforestation, soil erosion, and reduced water retention. Additionally, over-extraction of water for agriculture, industry, and mining, along with excessive groundwater use, has lowered the water table in many areas of the basin. Limited irrigation coverage is another factor, as irrigation does not reach many upland agricultural areas despite the presence of reservoirs and canals. Lastly, climate change is exacerbating the frequency and severity of droughts through rising temperatures and erratic rainfall patterns. Drought-prone districts such as Hazaribagh, Bokaro, Dhanbad, and Giridih in the upper basin of Jharkhand, along with Purulia and Bankura districts in West Bengal, also experience significant water stress. The impacts of drought extend to agriculture, water scarcity, energy production, and livelihoods, while also stressing the ecosystems in the basin area.

2.4 River Pollution

Water pollution, both surface and subsurface, is the main environmental issue in India (Hoque et al., 2022a). The water resource is being polluted due to various point and non-point sources of pollution (Hoque & Islam 2024). The sources of water pollution in India are untreated municipal sewage, discharge of industrial effluents (Sarkar et al., 2021), and intensive use of agrochemicals and agricultural runoff (Hoque et al., 2022b).

In the upstream area of Panchet Dam, the river Damodar, when it flows through the coal belt of Jharkhand, gets seriously polluted due to the discharge of industrial effluents from fertilizer factories, cement factories, coal washeries, mica factories, steel plants, etc. The National Environmental Engineering Research Institute (NEERI) prepared a pollution report of the Damodar in 1994. In general, the rivers of West Bengal are said to be polluted if the river water is polluted with organics (BOD >3 mg/L). It was found that between Santhaldih and Durgapur Thermal Power Station (DTPS), the BOD level of river water exceeds the critical limit (i.e., > 3 mg/L) (Table 2.11). The NEERI reported significant results at selected locations of the study area presented in Table 2.11.

Table 2.11 Critical biological oxygen demand status of Damodar River between Santhaldih and Durgapur Thermal Power Station (West Bengal)

Sl. No.	Pollutant status (mg/L)	Sites	Reason
1	BOD = 3.2	Confluence of Kadamda *Nallah* and Damodar	Effluent from Bhojudih coal washeries and Santhaldih Thermal Power Station (STPS)
2	BOD = 5.2	Damodar at Raniganj	Discharge of domestic waste from the township
3	BOD = 5.6	Confluence of Tamla *Nallah* and Damodar	Effluents from steel plants and chemical industries
4	BOD = 5.0	DTPS outfall	Effluent discharge from the power station

Source: Patra (2008)

Table 2.12 Key effluents entering the Damodar River in selected locations between Burnpur and Durgapur (West Bengal)

Sl. No.	Sites	TDS (mg/L)	TSS (mg/L)	pH	DO (mg/L)	BOD (mg/L)	COD (mg/L)	NH_4^+ N (mg/L)
1	Drain from Burnpur IISCO & Township	600	19	7.7	4.3	28	108	3
2	Nunia *Nallah* before Damodar	540	29	7.9	6.4	5	48	8
3	Singaran *Nallah* before Damodar	480	19	8.2	7.9	26	160	0
4	Outlet of DTPS Ash Pond	180	1889	8.0	7.1	4	52	1.2
5	Tamla *Nallah*-canal siphon crossing	300	662	8.5	5.2	48	360	8
6	Ningha colliery near Raniganj	900	37	7.8	1.9	22	90	2
7	Ricket & Colman Ltd, Asansol	780	9	7.8	7.7	3	20	0.5
8	DTPS colony discharge, Durgapur	240	12	7.6	1.0	30	224	–

Source: Patra (2008)

Table 2.12 shows that DO ranges from 1.0 to 7.9 mg/L, which exceeds CPCB (Central Pollution Control Board, India) standards (4 mg/L) and BOD ranges from 3 to 48 mg/L, which also exceeds CPCB standards (3 mg/L). In most of the sites, the free ammonia is found to exceed the value of 1.2 mg/L (critical). According to the CPCB standard, the water of the Damodar is categorized as 'Water Class D' (propagation of wildlife and fisheries only). In the river, pH values ranged from 7.5 to 8.9 during the monsoon and 7.7–8.8 during the post-monsoon season (Ghosh & Banerjee, 2012). The pH in the water samples exceeds the Food and Agricultural Organization (FAO) standards (6.5–8.0) for agricultural application but is within the recommended Bureau of Indian Standards (BIS) irrigation limiting standards (5.5–9.0). Lead (Pb) is an extremely pervasive and toxic environmental contaminant in water, and its values range from 0 to 0.071 mg/L during monsoon and 0–0.092 mg/L during post-monsoon (Ghosh & Banerjee, 2012). The concentrations of Pb in the river are within the recommended FAO standards (5.0 mg/L) for agricultural application. Iron ranged from 0.12 to 1.52 mg/L during monsoon and 0.052–3.55 mg/L during post-monsoon. It is well within FAO standards (5.0 mg/L), but exceeds BIS irrigation standards (3 mg/L) (Ghosh & Banerjee, 2012). The presence of heavy metal pollutants and the concentration of metals are in the order of Fe > Mn > Pb.

There are several coal washeries, coke-oven, and softcoke plants on both banks of the Damodar River. The traditional industries, chiefly supported by coal, iron, and steel have undergone a rapid diversification and new industrial ventures, which include mainly heavy engineering, fertilizers, coal-based chemicals, current plants, and thermal power generation, are making a dominant impact on the water resources of the Damodar. The major industries have their treatment facilities, but they have discharged many pollutants into the river system. Those industries having a direct impact on Damodar are identified as (1) Santaldih Thermal Power Station (STPS), (2) Bhojudih Coal Washeries, (3) Dishergarh Thermal Power Station, (4) Indian Iron and Steel Company Ltd. (IISCO) Burnpur, (5) Bengal Paper Mill Company Ltd. Raniganj, (6) Alloy Steel Plant (ASP) Durgapur, (7) Durgapur Steel Plant (DSP), (8) Durgapur Chemicals Ltd. (DCL), (9) Durgapur Thermal Power Stations (DTPS), and (10) Hindustan Fertilizer Corporation Ltd. (HFC) Durgapur (Patra, 2008) (Table 2.13).

De et al. (1980, 1985) studied the impact of industrial effluents on the Damodar River near Durgapur and found that Tamla *Nallah*, which carried the waste of major industries like Durgapur Projects Limited (DPL) and DCL, constantly contaminated the river and the barrage. High values (beyond the limit of BIS) of mixed oxides, ammonia, calcium, magnesium, total water-soluble exchange cation, and anion and COD were noted in the Bengal Paper Mill, Singaran *Nallah*, and HFC drain confluence, but maximum values were found at Tamla *Nallah* confluence where the pollution load from Durgapur region appeared to be maximum (De et al., 1985). The levels of toxic elements (e.g., arsenic, cadmium, chromium, lead, and zinc) were significant at 30 and 40 km downstream of the river from the Bengal Paper Mill, Raniganj (De et al., 1985). A similar study was done by De (1993) to monitor the effects of pollution on aquatic plants and paddy husks in the Asansol-Durgapur industrial belt. He found that the Asansol-Durgapur region contaminated the barrage water, while Tamla *Nallah* and the industrial drains of the Durgapur region contaminated the river directly. River water is not safe for drinking purposes as it has lost its self-purification capacity. Damodar River sediments, between Raniganj and Panagarh, were polluted with toxic elements (Cd, Cr, Hg, Pb, Zn, Ni, etc.) and the water is unfit for domestic purposes. Patra (2008) monitored the river and Tamla *Nallah*, which had COD in the range of 40–80 mg/L; BOD of 11–25 mg/L and TSS as 50–160 mg/L. Fluvial sediments of the river, downstream of the confluence of Tamla *Nallah*, had a maximum iron content of about 21,000 mg/kg, zinc 45 mg/kg, copper 16 mg/kg, total chromium 28 mg/kg, nickel 17 mg/kg, and lead as 5 mg/kg (Patra, 2008). Moreover, municipal waste is an important source of water pollution.

Table 2.13 Identified toxic effluents of major industries entering the Damodar River systems

Sl. No.	Industry	Major Pollutants	Receiving water body (impact)	Type of treatment available (adequacy of facilities)
1	Santaldih Thermal Power Station	TSS, O & G	Kadamda *Nallah* (significant)	Settling tank (inadequate)
2	Bhojudih Coal Washeries	TSS, TDS, COD	Damodar River (significant)	Settling tank (inadequate)
3	Dishergarh Thermal Power Station	TSS	Damodar River (insignificant)	Settling tank (adequate)
4	Indian Iron & Steel Co. Ltd., Burnpur	Fe, TSS, O & G	Damodar River (significant)	ETP with lagoons (inadequate)
5	Bengal Paper Mill Co. Ltd., Raniganj	TSS, BOD, COD	Damodar River (significant)	Only lagoon (inadequate)
6	Alloy Steel Plant, Durgapur	pH, TSS, COD, O & G, Cr, Ni	Tamla *Nallah* (significant)	Sludge pit, Pre-neutralization with settler (adequate)
7	Durgapur Steel Plant	TSS, BOD, COD, O & G, Phenol, Cyanide	Tamla *Nallah* (significant)	BOD plant, settling ponds, ash ponds, oil catch pit (adequate)
8	Durgapur Chemicals Ltd.	TSS, BOD, Phenol, COD, O & G	Tamla *Nallah* (significant)	Hg controls (inadequate)
9	Durgapur Thermal Power Stations	TSS, BOD, COD, O & G	Damodar River	Ash ponds (inadequate)
10	Hindustan Fertilizer Corporation Ltd. (HFC), Durgapur	Cr, As, NH_4^+–N	Tamla *Nallah* (significant)	Removal plant for As and Cr (inadequate)

Source: Patra (2008)

The water is polluted due to the urban disposal and their effluents are discharged into the river through the canal and tributaries. In the basin area, there are 49 (CIFRI, 1988) such sources, and their huge amount of effluents with pollutants mix into the river. Most of the town is an industrial or mining-based town in the basin area. According to CIFRI (1988), the river from Bermo to Bardhaman receives 273.134 MLD of municipal effluent with 174.43 tons of BOD load per day. Wastewater from industrial complexes and livestock farms contributes more NH_4^+–N, COD, Cu, and Pb into the river.

2.5 DVC and Dams in the Context of Hydrological Hazards Management

Different strategies have been taken by the government authorities and local people to manage water resources in the basin area, such as the construction of embankments, weirs, dams, and barrages. The central government consulted with the state governments (Bihar and West Bengal) and conceived the Damodar Valley project (DVP) to overcome floods and inundation problems, to utilize the river water for hydroelectricity generation, irrigation, and industries, and to develop the region agriculturally and industrially. Later on, it is called the Damodar Valley Corporation (DVC) along the lines of the Tennessee Valley Corporation of the USA. One of the leading Tennessee Valley Corporation engineers, W.L. Voorduin, prepared the preliminary memorandum that presented a flood control, power generation, irrigation, and navigation plan in the Damodar Valley. Finally, the DVC was instituted through the Parliament Act and was assigned Rs. 131.71 crores for the project implementation. Initially, this project was designed to construct seven dams, namely, (i) Maithon, (ii) Tilaiya, (iii) Balpahari (all on Barakar River), (iv) Panchet Hill, (v) Aiyar (on Damodar River), and one across (vi) Konar, and one on (vii) Bokaro. These dams had 4.7 million acre-feet total storage capacity. Thus 4.7 million acre feet total storage capacity created was considered sufficient to supply regular water for irrigation during the entire non-monsoonal period of 8 months. The Damodar, a seasonal river was thus converted into a perennial river to ensure irrigation throughout the year. In addition, three major hydropower plants, one each at Tilaiya, Maithon, and Panchet Hill, with a total installed capacity of 104,000 Kilowatt (KW) hydropower, were planned for power generation. A thermal power plant was also planned with an installed capacity of 150,000 KW thermal power at Bokaro.

The DVC is a multitasking organization and was established based on the Tennessee Valley Authority (TVA) model on 7th July 1948 by the Indian Constituent Assembly Act (Act No. XIV of 1948). It is independent India's first multipurpose river valley project, also known as the Damodar Valley Project (DVP). Under the scheme of DVC, eight (8) dams had been proposed to be constructed on the upper basin area of the Damodar River, but having lack of funds, only four (4) dams, like Maithon and Tilaiya on the river Barakar, Panchet on the river Damodar and Konar on the river Konar, have been constructed in the basin area. India's first multipurpose dams Tilaiya in 1953, Konar in 1955, Mithan in 1957, and Panchet in 1959, were installed by DVC (Table 2.S1). During this period, one barrage, i.e., Durgapur, was installed on the river Damodar at Durgapur in 1955 to provide water for irrigation purposes. Later, one more dam, i.e. Tenughat, was constructed on the river Damodar at Tenughat in 1978. The broad objectives of this organization are to develop and manage the natural resources of the DRB for regional and socio-economic development. These dams have been constructed mainly to control floods, for irrigation, and hydroelectricity generation purposes under the scheme of DVC.

2.5.1 Flood Control

Before the construction of the dams, serious floods happened in the basin area as mentioned in the earlier section. Hence, controlling floods is one of the major objectives of the DVC dams as several devastating floods occurred in the valley region. The valley region comes under the south-west monsoon of the Indian climate that hits during the period from the 4th week of July to the 1st week of August, causing heavy rainfall in the catchment area. The normal average rainfall occurs at 1335 mm in the Damodar catchment area (Ghosh et al., 2016).

The original plan of DVC proposed seven reservoirs with a 2.915 million acre-feet flood storage capacity in the basin (DVC, 2022). In the first stage, four multi-purpose dams, such as Panchet, Maithon, Konar, and Tilaiya were planned for a 1.51 million-acre feet floodwater capacity reservoir. However, due to a lack of land acquisition, it gained 1.047 million-acre feet of flood reserve capacity (DVC, 2022). The creation of these reservoirs reduces flood probability in the Damodar Valley region by managing the Damodar River water during the monsoon period. DVC reduces flood probability substantially as it created 1292 million cu. M flood reserve and 1.85 million cu. M water storage capacity in the basin area (Damodar Valley Corporation, 2021). Floods basically occur in the lower Damodar basin. Being the terminal reservoirs, Panchet and Maithon have a great role in controlling these floods in the lower basin area. Panchet and Maithon reservoirs have a total of 542.76 MCM and 1086.76 MCM of floodwater reserve capacity (Fig. 2.12). Panchet has the highest floodwater reserve capacity (1086.76 MCM), and Konar has the lowest floodwater reserve capacity (55.51 MCM). The total flood reserve capacity under the four dams' reservoirs of DVC is 1862.66 MCM.

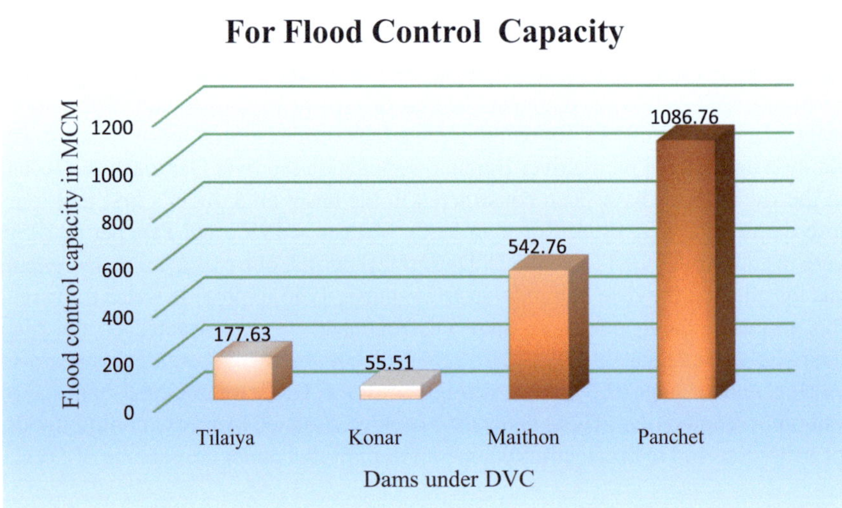

Fig. 2.12 Flood control capacity of DVC dams

2.5.2 *Flood Moderation*

The DVC dams have created a 2.5–6.51 lakh cusecs flood moderation capability in the basin area. This creation has played a significant role in controlling floods in the basin area, for example, peak floods (>10,000 m^3/s) happened five times at Rhondia in the pre-dam period (1933–1957), but it has been reduced substantially due to the moderation of floods by the DVC reservoirs in the post-dam period (Bhattacharyya, 2011). After the DVC dams, a combined peak inflow (>10,000 m^3/s) happened 15 times in the Maithon and Panchet dams (Table 2.14).

However, the downstream (from Maithon and Panchet dams) channel capacity of the Damodar River is 7085 m^3/s (Fig. 2.13). During 16–17 September 1958, peak inflow into the Maithon and Panchet reservoirs was 15,717.60 m^3/s and outflow was 4956 m^3/s and flood moderation was achieved by 10,761 m^3/s (68.46%). The highest combined inflow (21,919.68 m^3/s) into the Maithon and Panchet reservoirs was recorded on 27 September 1978. This inflow was moderated to 4616.16 m^3/s combined outflow and 17,303.52 m^3/s (78.94%) flood moderation achieved by these reservoirs. The average 63.74% flood moderation was achieved during 1958–2007 in the post-dam period by combined Maithon and Panchyet due to flood management by DVC dams (Table 2.14). These reservoirs have achieved 32–80% of flood

Table 2.14 Flood moderation by Maithon and Panchet dams

Date	Max. inflow (m^3/s)	Max. outflow (m^3/s)	Flood moderation achieved (m^3/s)	Flood moderation achieved (%)
16–17 September, 1958	15,717.60	4956	10,761.00	68.46
1–2 October, 1959	17,646.36	8156.16	9487	53.76
27 September, 1960	9855.36	2605.44	7249.92	73.56
2–3 October, 1961	14,643.12	4559.52	10,053.6	68.66
2–3 October, 1963	12,772.32	3426.72	9345.6	73.17
24–25 October, 1963	13,168.8	2577.12	10,591.68	80.43
29–30 July, 1964	10,563.36	2208.96	8354.4	79.09
3–4 September, 1970	8269.44	2265.6	6003.84	72.60
16–18 July, 1971	12,007.68	5125.92	6881.76	57.31
12–13 October, 1973	16,652.16	4956	11,696.16	70.24
26–27 September, 1975	9742.08	3143.52	6598.56	67.73
27 September, 1978	21,919.68	4616.16	17,303.52	78.94
27–28 August, 1980	9657.12	4219.68	5437.44	56.30
25 June, 1984	10,600	4800	5800	54.72
14–17 September, 1993	7000	2800	4200	60.00
29 September, 1995	17,360.16	7080	10,280.16	59.22
25 September, 1999	10,300	3400	6900	66.99
19–21 August, 2000	10,800	5700	5300	49.07
24 September, 2006	14,400	6900	7500	52.08
25 September, 2007	11,100	7500	3600	32.43

Source: Based on Bhattacharyya (2011); Ghosh and Guchhait (2014)

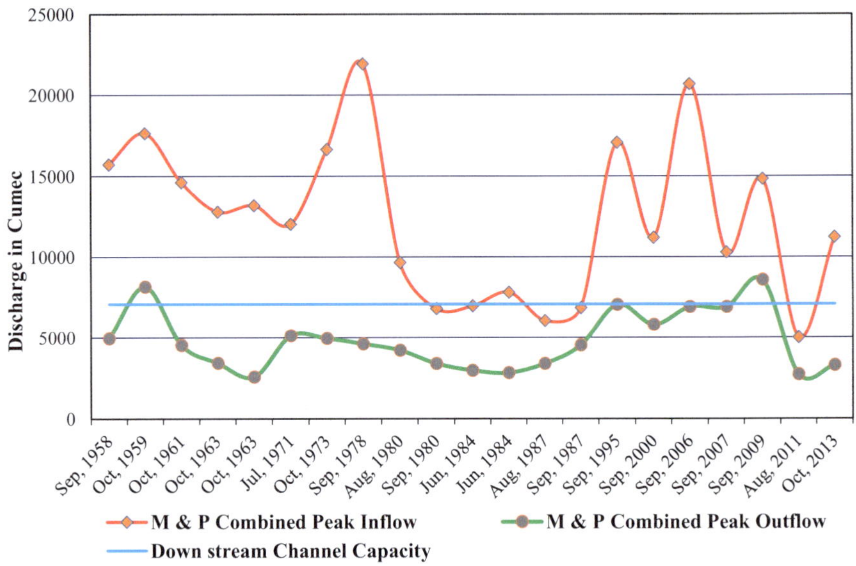

Fig. 2.13 Flood moderation at the downstream of Maithon (M) and Panchet (P) dams

moderation after the DVC dams. The highest (80.43%) flood moderation was achieved during 24–25 October 1963.

2.5.3 *Critical Outlook on Flood Geomorphological Processes*

After the devastating flood of 2000, the state experienced another severe flood in 2021. About 22 lakh people across eight districts (*Purba* and *Paschim* Midnapore, Bankura, Hooghly, Birbhum, Howrah, *Purba*, and *Paschim* Barddhaman) were affected. The lower Damodar River Basin (some parts of *Purba* and *Paschim* Barddhaman, Bankura, Hooghly, and Howrah districts) was also badly impacted. Udaynarayanpur and Amta 2 blocks are the most severely affected areas within the Damodar River basin. On the midnight of 30th September, the embankment of the Damodar River was breached, allowing water to flow into villages in Udaynarayanpur, Howrah. The Damodar River breached in approximately 12 locations, with around 110 m of embankment breached in five spots at Ghola, 20 m at Bhabanipur, 21 m at Shibanipur, 42 m in two locations of Jaynagar, and 35 m in two locations of Akna Dwittia, while an 18-m breach occurred in Maja Damodar (Irrigation and Waterways, 2021). Water from the Damodar River entered villages from 15 river sites. By the evening of 1st October, 100 villages across 9 gram panchayats in Udaynarayanpur were submerged, and from that afternoon, water began to inundate Amta 2 block, eventually covering nearly the entire area by night. Fourteen gram panchayats in the Amta 2 block were entirely submerged. Udaynarayanpur village was completely isolated from Friday morning, as the transport system collapsed; water levels reached throat height on the Udaynarayanpur—Dihirentshut road. In addition to roads and shops, water entered the police station and the Block Development Office (BDO) office in Udaynarayanpur. Around 16,000 victims were relocated to approximately 92 flood shelter camps. Ridhi Khara, a 2.5-year-old boy, was washed away by the Damodar River flood while his parents were having lunch in Shibpur village of Kurchi Shibpur gram panchayat. Some parts of Udaynarayanpur and Amta 2 block are submerged under 3 m of floodwater. Pursura block and 24 gram panchayats of Khanakul, two blocks in Hooghly district, are completely inundated. Shanpur, Chila Dangi, Pursura I, and Srirampur panchayats are severely impacted. On 2nd October, Kaif Mullick from Sathpaithya village of Khanakul was washed away. Blockura is heavily affected by the flows of the Damodar, Bodai, and Shali Rivers. Some villages in Sonamukhi and Patrasayar blocks have been flooded. About 60,000 people from 230 gram panchayats have been affected, and around 4000 buildings have been damaged. Two women died following the collapse of a mud wall; one died of snakebite while crossing a waterlogged road in Bhamira village, and another was washed away in the floodwaters. Approximately 8000 people are sheltered in 65 flood camps, with 125,000 acres of paddy land submerged. 13,000 victims are sheltered in 116 flood camps in West Barddhaman, where two people died from collapsing walls and one from electric shock. One person was washed away during a fishing expedition in East Barddhaman. Five columns of the Indian Army are engaged in rescue operations to assist those affected. This depicts the devastating scenario caused by the flood.

Now, let us examine the reasons behind this flood. The flood is a major political issue, and the West Bengal Government has blamed DVC for this incident, which the West Bengal Chief Minister (CM), Mamata Banerjee, has described as a 'man-made flood'. Ms. Banerjee asserts that monsoon rainfall is not responsible, but rather the large water releases by DVC without proper prior notice to West Bengal, coinciding with heavy rainfall in Jharkhand, have contributed to the flooding in West Bengal. According to chief secretary Harikrishna Dwivedi, in July 2021, the state also experienced flooding due to excessive rainfall and water releases, and as of 30th September, many areas are still struggling to recover, with some remaining submerged. In this critical situation, DVC released a substantial volume of water, starting with 49,000 cusecs from Panchet and Maithon at noon on 30th September, which increased to 1 lakh cusecs at 1 pm on the same day, then to 1.25 lakh cusecs at 8:30 pm, and further to 1.5 lakh cusecs at 8 am on 1st October 2023. Additionally, at 3 am on 30th September, 80,000 cusecs of water were released from the Sikatia dam in Jharkhand, which increased to 1.2 lakh cusecs by 9:45 am on 1st October. Combining all sources, approximately 10 lakh cusecs of water were released by DVC and Jharkhand over 2 days. DVC dismissed Banerjee's criticism, stating that the decision to release water was made with the advice of the Damodar Valley Reservoir Regulation Committee (DVRRC), headed by a member of the Central Water Commission, with representatives from West Bengal and Jharkhand present. They further stated that if DVC releases more than 23,000 cusecs of water, they will inform the irrigation department of West Bengal, with the initial warning given 6 h before the water release. DVC also added that flood warning messages were effectively communicated to the chief engineers of the West Bengal Government, district magistrates of *Purba* and *Paschim* Bardhaman, Bankura, Purulia, as well as superintending and executive engineers of Durgapur. From there, the executive engineer of Durgapur relays information to the district magistrates of Howrah, Hooghly, SDOs, ADMs, BDOs, and others. Regarding DVC's warnings, the state claims to have received notification from DVC that they were compelled to release 1.5 lakh cusecs of water from Panchet and Maithon on 30th September. State sources also mentioned that in the initial warning email, DVC did not specify the exact volume of water; however, 2 h before the release, the CWC chairman sent an email detailing the amount, which was forwarded to the ADM of Asansol and West Barddhaman authorities so they could monitor the barrage. Despite DVC's actions, the CM also suggested that the Jharkhand Government should renovate the dams and proposed joint planning with West Bengal.

Apart from the political issue, if we focus on this matter, we notice that the rainfall statistics of the affected districts, such as the normal rainfall of Bankura, East and West Barddhaman, and Hooghly, are 466 mm, 361.1 mm, 318.3 mm, respectively, while the actual rainfall amounts are 249 mm, 225.7 mm, and 240.1 mm, respectively. Therefore, the deviation from normal to actual rainfall is not significant enough to account for such a massive flood. While the water released from DVC is substantial, it causes a colossal flood. When dams and barrages were constructed, there was sufficient capacity to store an adequate amount of water in the reservoirs. However, in subsequent years, the storage capacity declined considerably due to sedimentation in the reservoirs. Therefore, desilting at the appropriate times is important for reducing severe floods in West Bengal. The Bengal Irrigation Department also proposed that constructing new dams upstream of the Maithan and Panchet dams within the DVC network, rather than desilting existing dams, could help prevent regular flooding in south Bengal caused by water releases.

This chapter clearly demonstrates the unique facets of the study region with special reference to the hydrological extremes in the form of floods and droughts. This establishes a linkage of why environmental flow analysis from the quantitative perspective is required. Thus, the quantitative dimension of the environmental flow of the Damodar River is discussed in Chap. 3.

References

Abdulkareem, J. H., Sulaiman, W. N. A., Pradhan, B., & Jamil, N. R. (2018). Relationship between design floods and land use land cover (LULC) changes in a tropical complex catchment. *Arabian Journal of Geosciences, 11*(14), 376. https://doi.org/10.1007/s12517-018-3702-4

Bhattacharyya, K. (2011). *The Lower Damodar River, India: Understanding the human role in changing fluvial environment*. Springer.

Bhattacharyya, K., & Wiley, M. J. (2014). Dams, riparian settlement and the threat of climate change in a dynamic fluvial environment: A case study of the Damodar River, India. In *Large*

dams in Asia: Contested environments between technological hydroscapes and social resistance (pp. 75–100).
Census of India (2001). District Census Handbooks, Directorate of Census Operation, India. (Accessed 15 March 2019).
Census of India (2011). District Census Handbooks, Directorate of Census Operation. India. (Accessed 15 March 2019).
Church, M. (2015). Channel stability: Morphodynamics and the morphology of rivers. In *Rivers–physical, fluvial and environmental processes* (pp. 281–321). Springer. https://doi.org/10.1007/978-3-319-17719-9_12
Damodar Valley Corporation. (2021). *DVC Reservoir Overview*. www.dvc.gov.in/dvcwebsite_newl/dams-barrages/DamodarValleyCoporation
DVC. (2022). *Annual Report 2021–22*. Retrieved December 18, 2023, from https://www.dvc.gov.in/dvcwe bsite_new1/downloads/annual-report/
Ghosh, S. (2014). The impact of the Damodar valley project on the environmental sustainability of the lower Damodar basin in West Bengal, Eastern India. *OIDA International Journal of Sustainable Development, 7*(02), 47–54.
Ghosh, S., & Guchhait, S. K. (2014). Hydrogeomorphic variability due to dam constructions and emerging problems: A case study of Damodar River, West Bengal, India. *Environment, Development and Sustainability, 16*, 769–796. https://doi.org/10.1007/s10668-013-9494-5
Ghosh, P. K., Bandyopadhyay, S., Jana, N. C., & Mukhopadhyay, R. (2016). Sand quarrying activities in an alluvial reach of Damodar River, Eastern India: Towards a geomorphic assessment. *International Journal of River Basin Management, 14*(4), 477–489. https://doi.org/10.1080/15715124.2016.1209509
Hunter, W. W. (1876). *A statistical account of Bengal* (Vol. 4). Trübner & Company.
Khodaei, H., Nasiri Saleh, F., Nobakht Dalir, A., & Zarei, E. (2025). Future flood susceptibility mapping under climate and land use change. *Scientific Reports, 15*(1), 12394. https://doi.org/10.1038/s41598-025-97008-0
Patel, M., & Kumar, B. (2017). Flow and bedform dynamics in an alluvial channel with downward seepage. *Catena, 158*, 219–234. https://doi.org/10.1016/j.catena.2017.07.009
Sanyal, J., & Lu, X. X. (2004). Application of remote sensing in flood management with special reference to monsoon Asia: A review. *Natural Hazards, 33*, 283–301.
Sarkar, B., Islam, A., & Das, B. C. (2021). Role of declining discharge and water pollution on habitat suitability of fish community in the Mathabhanga-Churni River, India. *Journal of Cleaner Production, 326*, 129426. https://doi.org/10.1016/j.jclepro.2021.129426
Vázquez-Tarrío, D., Ruiz-Villanueva, V., Garrote, J., Benito, G., Calle, M., Lucía, A., & Díez-Herrero, A. (2024). Effects of sediment transport on flood hazards: Lessons learned and remaining challenges. *Geomorphology, 446*, 108976. https://doi.org/10.1016/j.geomorph.2023.108976
Patra, S. (2008). Pollution profile of Damodar River in a particular stretch. Unpublished Report of Master of Technology in Conservation of Rivers and Lakes, Alternate Hydro Energy Centre, Indian Institute of Technology Roorkee. http://shodhbhagirathi.iitr.ac.in:8081/jspui/bitstream/123456789/12088/1/HYDG13946.Pdf
Banerjee, K., Banerjee, R., Sen, A., Mukhopadhyay, S., Ach, S., Singh, D., & Mandal, B. (2003). Damodar river pollution and health hazards. Journal of the Indian Medical Association, 101(2), 104–106.
Dutta, T., Chaudhuri, H., & Maji, C. (2021). Water Pollution in Damodar River Basin—A Statistical Analysis. In Advances in Water Resources Management for Sustainable Use (pp. 187–215). Springer, Singapore. https://doi.org/10.1007/978-981-33-6412-7_15
Majumder, M., Roy, P., & Mazumdar, A. (2010). An introduction and current trends of Damodar and Rupnarayan River Network. Impact of climate change on natural resource management, 461-480. https://doi.org/10.1007/978-90-481-3581-3_25
Tiwary, R. K., & Dhar, B. B. (1994). Environmental pollution from coal mining activities in Damodar river basin, India. Mine water and the environment, 13(1), 1–10.

Singh, A. K. (1997). Major and trace elements geochemistry of Damodar River Basin. Jawaharlal Nehru University.

DVC (1992). Damodar Valley - evolution of the grand design. https://www.dvc.gov.in/storage/app/impLink_meida/66Damodar-ValleyEvolution-of-the-Grand-Design_Compressed.pdf

Geological Map of Jharkhand (2015). Mines and Geology. https://geology.jharkhand.gov.in/documents/Geological%20Map%20of%20Jharkhand.pdf

Mondal, G. C., Singh, A. K., & Singh, T. B. (2018). Damodar river basin: storehouse of Indian coal. The Indian Rivers: Scientific and Socio-economic Aspects, 259–272. https://doi.org/10.1007/978-981-10-2984-4_21

Banerjee, U.S (2013). Studies on environmental geochemistry of river Damodar along the stretch of Dishergarh to Burdwan, West Bengal, India. The University of Burdwan.

Prasad, N., & Giri, M. M. (2009). Hydrogeological Characteristics of the Jamuniya Basin, Jharkhand. Earth System Sciences: Felicitation Volumes in Honour of Professor VK Verma, 1, 254.

Mahadevan, T. M. (2002). Text book series 14: Geology of Bihar and Jharkhand. *Geological Society of India, 60*(4), 380–380.

Ghose, N. C. (1983). Geology, tectonics and evolution of the Chhotanagpur granite-gneiss complex, Eastern India. In Structure & tectonics of Precambrian rocks of India (pp. 211–247). Hindustan Publishing Corporation, Delhi

Ghosh, S. (2011). Hydrological changes and their impact on fluvial environment of the lower damodar basin over a period of fifty years of damming The Mighty Damodar River in Eastern India. *Procedia-Social and Behavioral Sciences, 19*, 511–519. https://doi.org/10.1016/j.sbspro.2011.05.163

Ghosh, S., Hoque, M. M., Saha, U. D., & Islam, A. (2022). Assessment of dam-induced changes in ecogeomorphological behaviour and fluvial functionality in the Damodar River, West Bengal, India. *AQUA—Water Infrastructure, Ecosystems and Society, 71*(6), 722–750. https://doi.org/10.21203/rs.3.rs-1002245/v1

Sardar, B., Singh, A. K., Raghuwanshi, N. S., & Chatterjee, C. (2014). Hydrological modeling to identify and manage critical erosion-prone areas for improving reservoir life: case study of Barakar Basin. *Journal of Hydrologic Engineering, 19*(1), 196–204. https://doi.org/10.1061/(ASCE)HE.1943-5584.0000749

Chandra, S. (2003). India: Flood management-Damodar river basin. World Meteorological Organization and Associated Programme on Flood Management, Integrated flood Management-Case study, 1–10.

Olutaiwo, A. O., & Olushola, L. A. (2017). Effect of banana leaf ash on cement-modified lateritic soil. *International Journal of Engineering and Technical Research, 7*(6), 264983.

Choudhury, S., Chattopadhyay, G. S., & Das, N. (2012). Damodar River Basin; Economic Development & Nature. In Sustainable Development: An Interdisciplinary Approach. Seminar sponsored by University Grants Commission, India.

Ghosh, A. R., & Banerjee, R. (2012). Qualitative evaluation of the Damodar River water flowing over the coal mines and industrial area. *International Journal of Scientific and Research Publication, 2*(10), 1–6.

De, A. K., Sen, A. K., & Modak, D. P. (1980). Some industrial effluents in Durgapur and their impact on the Damodar River. *Environment International, 4*(2), 101–105.

De, A. K., Sen, A. K., Karim, M. R., Irgolic, K. J., Chakraborty, D., & Stockton, R. A. (1985). Pollution profile of damodar river sediment in Raniganj-Durgapur industrial belt, West Bengal, India. *Environment International, 11*(5), 453–458.

De, A.K. (1993). Growth of water pollution in Asansol -- Durgapur industrial belt and its impact on the water quality of the Damodar River. Environmental Research, MOEF, India

CIFRI (1988). The River Damodar and Its Environment. Indian Council of Agricultural Research, Barrackpore, Bull. No. 73. http://www.cifri.res.in/Bulletins/Bulletin%20No.79.pdf

Hoque, M. M., Islam, A., & Ghosh, S. (2022a). Environmental flow in the context of dams and development with special reference to the Damodar Valley Project, India: a review. *Sustainable Water Resources Management, 8*(3), 62. https://doi.org/10.1007/s40899-022-00646-9

Hoque, M. M., Islam, A., & Mahammad, S. (2022b). Assessing the surface and bottom river water quality for drinking purpose and human health risk analysis: a study of Damodar River, India. *Arabian Journal of Geosciences, 15*(23), 1734. https://doi.org/10.1007/s12517-022-10987-6

Hoque, M. M., Islam, A., Islam, A. R. M. T., Pal, S. C., Mahammad, S., & Alam, E. (2023). Assessment of soil heavy metal pollution and associated ecological risk of agriculture dominated mid-channel bars in a subtropical river basin. *Scientific Reports, 13*(1), 11104. https://doi.org/10.1038/s41598-023-38058-0

Hoque, M. M., & Islam, A. (2024). Spatio-temporal dynamics of ecological, bacteriological, and overall water quality of the Damodar River, India. *Environmental Science and Pollution Research, 31*(12), 18465–18484. https://doi.org/10.1007/s11356-024-32185-5

Hoque, M. M., Islam, A., Sarkar, B., & Saha, U. D. (2024). Assessing the surface and bottom river water quality for irrigation: A study of Damodar River, India. *International Journal of Energy and Water Resources, 8*(4), 545–562. https://doi.org/10.1007/s42108-022-00206-z

Hoque, M. M., Islam, A., Pal, S. C., Sarkar, B., & Hossain, Z. M. (2025a). Assessment of fish habitat suitability in the wake of elevated pollution of a tropical river in India. *Environmental Science and Pollution Research, 32*(3), 1208–1222. https://doi.org/10.1007/s11356-024-35685-6

Hoque, M. M., Pal, S. C., Mandal, S., & Islam, A. (2025b). Assessment of mollusc habitat suitability of an anthropogenically altered tropical river in India. *Environmental Monitoring and Assessment, 197*(10), 1–18. https://doi.org/10.1007/s10661-025-14593-3

Islam, A., Sardar, N., Mohinuddin, S., Hoque, M. M., Sengupta, S., Das, B. C., ... & Sengupta, B. (2023). Quasi-equilibrium channel metamorphosis in planform of a subtropical river in India in post-dam period. *Catena, 221*, 106793. https://doi.org/10.1016/j.catena.2022.106793

Chapter 3
Damodar Valley Corporation and Quantitative Environmental Flow Alteration

3.1 Dams and Quantitative Flow Alterations in the Global and South East Asian Context

The concept of environmental flow has been developed to sustain and maintain freshwater and estuarine ecosystems (Hoque et al., 2022a). This concept has travelled across time and space all over the world (Blaney, 2013). The term 'Environmental flow' is defined as the quantity, timing and quality of water flow needed to sustain freshwater and estuarine ecosystems and the livelihoods of people who depend on these ecosystems (The Brisbane Declaration, 2007). It is the minimum flow required in an aquatic system to maintain its ecological integrity and health and to promote socio-economic sustainability. Generally, it refers to the river flow regime which controls all functions, processes, and structures of a river system and its floodplain and wetland (Poff et al., 2009). Besides, it also upholds the essential process of an ecosystem and maintains a healthy ecosystem (Smakhtin & Anputhas, 2006). It tries to balance the uses of river water between socio-economic and ecosystem requirements (Jain, 2015) or between social demands and ecological sustainability that impact the livelihood and ecosystem services (Hazra et al., 2015; Sarkar & Islam, 2022). Hence, its protection is crucial for conserving aquatic biodiversity in dammed rivers (Grantham et al., 2014). It is a management technique used to preserve ecosystem integrity and biodiversity (Karr, 1991; Silk & Ciruna, 2013). Its implementation is required to increase the river's potentiality, reduce ecosystem alteration, and conserve aquatic biodiversity (Grantham et al., 2014). It is significant for rivers as it sustains rivers by maintaining their ecological flow and is the biological engine of riverine ecosystems. It also allows a river to help clean up water, flush out salinity, improve fish habitat, encourage biodiversity, and provide mankind with ecotype services which are very important for us in terms of fresh and clean water provision (Chen & Wu, 2019).

M. M. Hoque, A. Islam, *Dams, Development and Environmental Flow in the Damodar River Basin*, https://doi.org/10.1007/978-3-032-11161-6_3

Thus, environmental flow assessment is vital nowadays because the hydrological conditions of a river system are being altered and diverted by dams (Ghosh et al, 2022). Moreover, it is also important due to the modification of stream flow and ecosystems through rising water extraction for various uses like agriculture, industry, hydropower generation, recreation, and domestic water supply (Verma et al., 2015). In this regard, several methodologies have been established for hydrological, hydraulic, water quality inputs, and biological, and socio-economic issues of the dammed river. Currently, 207 distinct environmental flow assessing methodologies exist in 44 countries in the world. The most significant are Hydrological, Hydraulic rating, Holistic method, Habitats simulation, (Tharme, 2003), and water quality and biological inputs approach (Tegos et al., 2018). In this context, several hydrological-based scholastic research works have been comparatively executed for different purposes all over the world, a few of them are given in Table 3.1. The major objective of the present investigation is to assess environmental flow using two hydrological methods, i.e., (i) Tennant and (ii) FDC in the DRB based on the daily, monthly, and annual discharge data obtained from different secondary sources.

3.2 Damodar River Flow Alterations

3.2.1 Flow Duration Curve

A flow duration curve (FDC) is used to design and measure environmental flows in river systems and to demonstrate a river's hydrological condition (Ghosh et al, 2022). It shows the percentage of time a stream's flow equals or exceeds a given value at a specific river section. This curve defines a minimum threshold value for maintaining and sustaining the ecological integrity of river systems. The concept of FDC emphasizes only the quantitative environmental flow, i.e. minimum flow associated with river health. The river's health problems occur due to low flow or no flow. From the concept of environmental flows, the flow regime includes the discharge of a specific magnitude, timing and frequency required to ensure a river system remains sustain socio-economically and environmentally. The Damodar River has five key components of flow regime, i.e., magnitude, frequency, rate of change, timing and duration, which are recognized as dominant factors for sustaining and maintaining biodiversity and ecological integrity. In FDC, on the horizontal axis (the probability axis), 17 points of percent, i.e., 0.01, 0.1, 1, 5, 10, 20, 30, 40, 50, 60, 70, 80, 90, 95, 99, 99.9, and 99.99 are used covering entire ranges of flow while on the vertical axis, a logarithm scale is used to flow discharge (Fig. 3.1). According to the FDC, any transformation of an FDC from the original or reference curve to the left indicate loss of flow (part of variability is lost) is as a consequence of decreased monthly flows assurance, i.e., the same flow will happen less frequently. This shift of FDC, e. flows is linked with environmental management classes (EMCs), used to demonstrate the hydrological conditions of a river (Jain &

Table 3.1 Hydrological methods-based comparative studies for different purposes

Used E. flows methods	Purpose	Reference
IFIM, Tennant, Wetted Perimeter, 7Q10 flow, Aquatic Base Flow (ABF)	In-stream flow methods most often used in North America	Reiser et al. (1989)
Tennant, Flow duration (Q95& Q90), Constant yield	In-stream flow methods used in Australia	Karim et al. (1995)
Tennant, 25% of the MAF, Monthly (Q50, Q90) flow, Aquatic Base Flow (ABF), 7Q10 flow	Comparisons of hydrologically based in-stream flow methods in 70 rivers of Atlantic Canada	Caissie and El-Jabi (1995)
Seasonal 7-day low flow, Seasonal 25% of mean flow, Seasonal Q80, Monthly Q50 & Q90	Examine the impact of temperature on low stream flow for 77 rivers in the Canadian Prairies and trends analysis	Yulianti and Burn (1998)
50-, 75-, 90-percentile FDC	Assessment to fulfil water requirement for an industrial plant located in the San and Brah River watershed, India	Pandey and Ramasastri (2003)
MAF, Q50, Q90, 7Q2, 7Q10	To protect the fish habitat	Caissie et al. (2007)
Default FDC	Assessment of EFRs in major Indian river basin	Smakhtin and Anputhas (2006)
FDC	Link EFRs with e. flows classes	Smakhtin and Eriyagama (2008)
FDC	Assess and design e. flows in the Brahmani-Baitarani River system, Odissa, India	Smakhtin et al. (2004)
Tennant	Assess optimal e. flows in Tungabhadra River, India	Jha et al. (2008)
Tennant, Q95 from FDC, Hydraulic	Computing the minimum water requirement to save biological activity in Safaood River, Iran	Babu and Kumara (2009)
Tennant, FDC shifting, Low-flow index, DRM, GEFC	Comparisons of hydrological-based methods for assessing EFRs to maintain basic functions in Shahr Chai River, Iran	Shokoohi and Hong (2011)

Source: Based on Verma et al. (2015)

Kumar, 2014). The expected condition of the river is called EMC. A higher EMC implies that more water is needed along with river water flow to maintain a river ecosystem.

The analysis of the FDC based on the annual peak flow at Rhondia in the Damodar River's pre-dam and post-dam periods reveals that the FDC has shifted left from the pre-dam period to the post-dam period (Fig. 3.2a), indicating that the variability is lost as a result of decreasing peak flows in the river systems. Besides, the daily flow analysis in the pre-dam (1940–1950) and post-dam periods (1993–2008) portrays the flow discharge of 0.7 m^3/s and 2 m^3/s, corresponding to 85% exceedance probability in the pre- and post-dam periods, respectively. However, the flow discharge

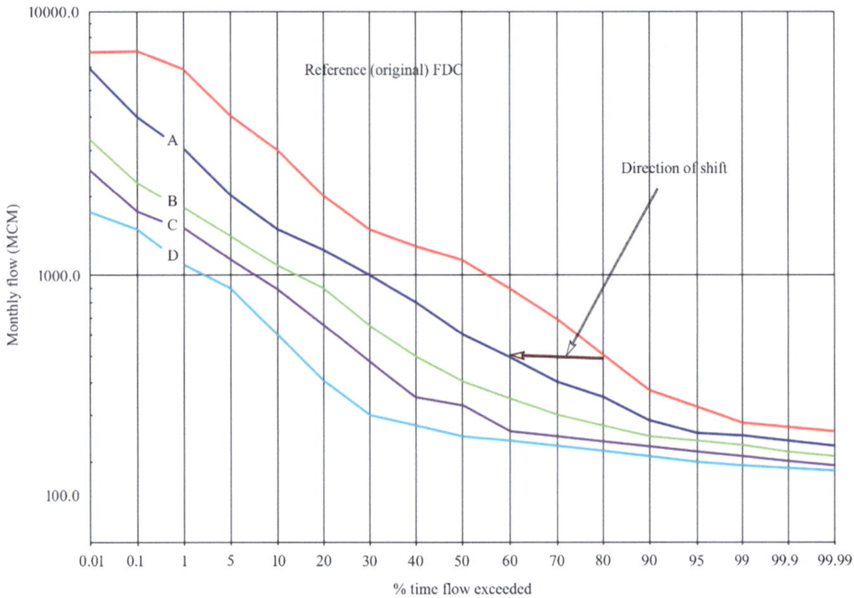

Fig. 3.1 Illustration of the estimation procedure of environmental flow duration curves for different environmental management classes (A, B, C, & D) based on Smakhtin and Anputhas (2006) and Jain and Kumar (2014)

of 3575 m^3/s and 2460 m^3/s corresponds to a 1% exceedance probability in the pre-dam and post-dam periods, respectively (Fig. 3.2b). The flow equivalent to <30% exceedance probability decreases and occurs less frequently, while the flow equivalent to >40% exceedance probability rises and occurs with high frequency due to flow regulation. The FDC of high flow (<30% exceedance probability) shifts left and low flow (>40% exceedance probability) shifts right after the DVC dams and barrage construction on the Damodar River (Fig. 3.2b). Thus, the analysis of FDC and the daily flow indicates that the river's flow variability is lost as a result of decreasing higher flow and increasing lower flow frequency in the river systems. Besides, at a 0.05 significance level and 1 degree of freedom, the results of ANOVA confirm a significant difference in peak discharge between the pre-dam and post-dam period ($p<0.05$, $F=72.40$ and $F_{crit}=3.95$). Hence, the daily flow variability has altered owing to flow regulation by DVC dams and barrages.

3.2.2 FDC in the Lean and Overall Period

This FDC (Fig. 3.3a, b) has been made based on the mean monthly discharge data of the overall (all months) and the lean period (December to May) at Rhondia in the Damodar River's pre-dam (1934–1957) and post-dam period (1958–1957).

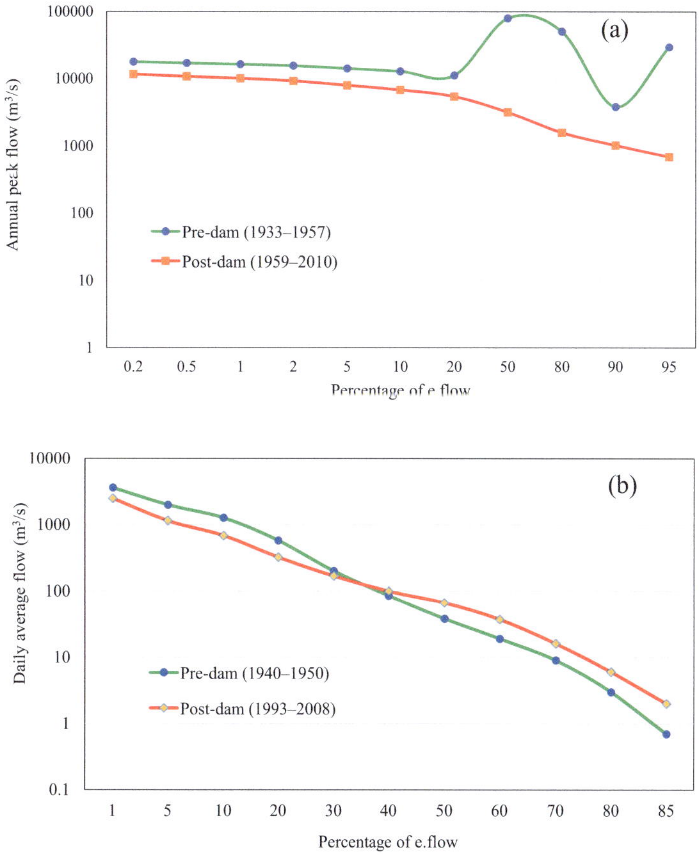

Fig. 3.2 Flow duration curve at Rhondia in pre-dam and post-dam periods of the Damodar River; (**a**) with respect to the annual peak flow and (**b**) with respect to daily average flow based on Bhattacharyya (2011)

According to this FDC, it is found that 98.9% and 97.8% exceedance probability is achieved for the overall and lean period, respectively, and their corresponding flow is 1 m^3/s in the pre-dam period. Hence, 1 m^3/s of flow was available in all the months of the pre-dam period. However, 99.4% and 98.8% exceedance probabilities are achieved for the overall and lean periods, respectively, in the post-dam with their corresponding flow of 8 m^3/s. Hence, 8 m^3/s of flow is available in all the months of the post-dam period. Thus, the monthly flow availability of the river at this section slightly (i.e., 1–8 m^3/s) increases after the DVC dams and barrage. Moreover, a 95% exceedance probability with its corresponding value of 16.45 m^3/s and 17.05 m^3/s is found for the pre- and post-dam periods, respectively. Hence, the environmental flow is found to be 1 m^3/s and 8 m^3/s in the pre- and post-dam correspondingly and the design environmental flow is 16.45 m^3/s and 17.05 m^3/s for the pre- and post-dam periods, respectively, at the Rhondia of the Damodar River.

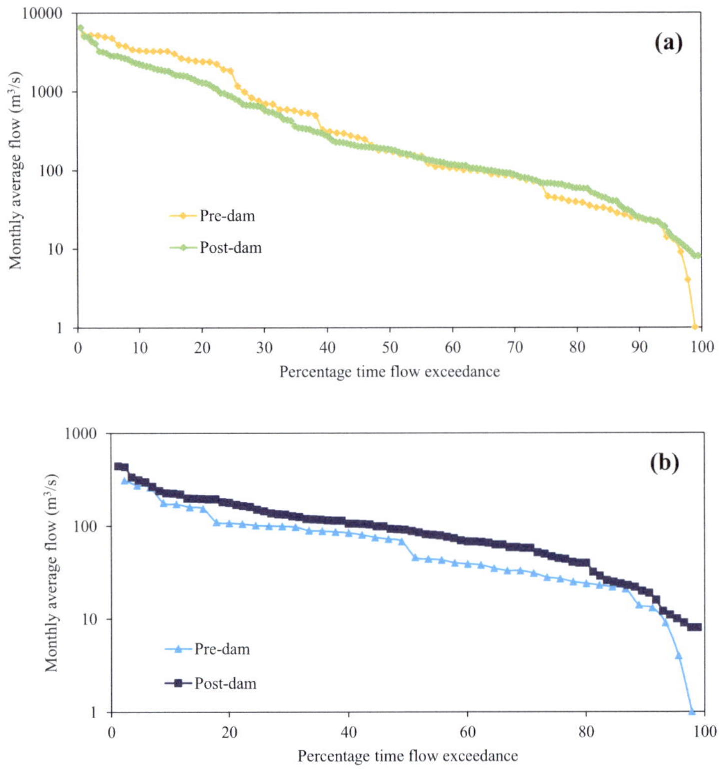

Fig. 3.3 Flow duration curve of 22 years of discharge data in the pre-dam and 42 years of discharge data in the post-dam period at Rhondia of Damodar River with respect to monthly average streamflow in the pre-and post-dam period; (**a**) overall and (**b**) lean period (December to May) based on Bhattacharyya (2011)

The downstream environmental flow variability is noticeable as a result of longitudinal disconnectivity by the installation of dams and barrages (Verma et al., 2015). After the fulfilment of large-scale industries and collieries' water requirements (Dhanbad-Durgapur region, nearly 115 km stretch along the Damodar River), the minimum flow in the lean period (mainly summer) varies between 1.14 m^3/s and 0.79 m^3/s based on the stochastic FDCs of 60-day daily mean flow. This flow is identified as the predictable e. flow of the lower Damodar River up to the Durgapur barrage. The intensive extraction of groundwater from the valley decreases the minimum base flow in the lean period, promoting the river bed dryness due to the aquifer's loss (deteriorating groundwater depth of >4 m below ground level in pre-monsoon between 2006 and 2015 in *Purba* and *Paschim* Barddhaman) that reduces lateral connectivity of the river with groundwater in summer to sustain flow. The survival of riverine flora and fauna is a serious issue under the water crisis, pollution, and climate change. India's National Green Tribunal (NGT) has directed maintaining a minimum or optimum e. flow of 15%–20% of the mean lean season

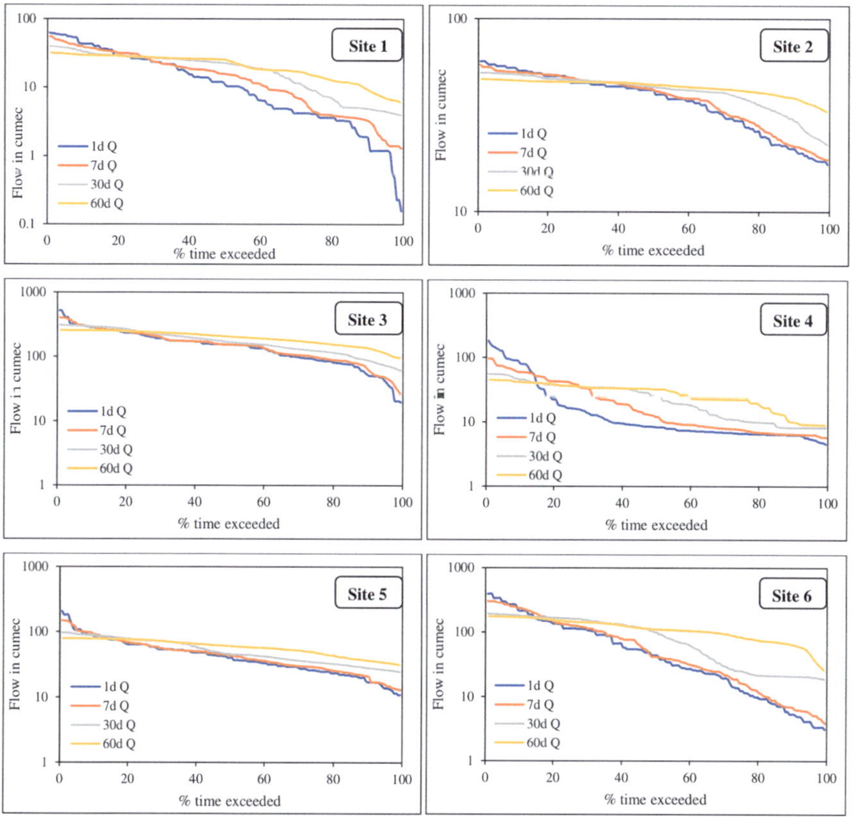

Fig. 3.4 Available period of record flow duration curves comparisons for 1-, 7-, 30-, and 60-day of six selected sites in the upper part of the Damodar River Basin. (Source: Water yearbook 2012, water resources department of Govt. of Jharkhand)

course in all states' rivers to make India's rivers free-flowing again. Smakhtin et al. (2007) assessed that a minimum flow of 10 m^3/s is required to maintain a highly polluted river's ecological balance in the Yamuna River Basin. The study portrays that the minimum flow ranges from 2.04 to 16.81 m^3/s (based on 7Q10 and 7Q100) in the monsoon-dominated and dam-induced DRB, which lies at a critical level concerning all Indian rivers.

3.2.3 FDC at Different Sites in Damodar River Basin

In the present study, FDCs have been constructed based on daily discharge data using 1-, 7-, 30-, and 60-day moving averages. The FDCs of 1-day, 7-day, and 30-day averages show no significant variations in FDC plots. Moreover, it is also observed that there is a higher probability of exceedance for 60-day average discharge, which is higher than 1-, 7-, and 30-day averages (Fig. 3.4).

Taking all the aspects, it is observed that a 7-day FDC is the most appropriate among the others. The FDCs also represented that the constructed FDCs differ from each site, which can be a variation in precipitation, geophysical setting, watershed conditions, and other meteorological reasons. It is also observed that the daily discharge is high at site 3, i.e., site-Tenughat dam downstream, while it is low at site 1, i.e., site-Tenughat Banaso Road, among the other sites.

3.2.4 Tenant Method

The e. flow concept mentions a minimum flow needed to maintain and sustain the aquatic ecosystem and ecological integrity. A minimum flow, i.e., 10% of the mean annual flow (MAF) is a prerequisite for aquatic habitat to maintain the biological integrity of the river ecosystem based on the Tennant method. The MAF was 1001.33 m^3/s in the pre-dam period at the Rhondia site of the Damodar River, but it reduced to 700.04 m^3/s in the post-dam period as a consequence of flow regulation by the DVC dams (Table 3.2).

Moreover, the optimum range of MAF was 600.80–1001.33 m^3/s in the pre-dam, whereas it is 420.02–700.04 m^3/s in the post-dam period at the Rhondia site of the Damodar River. The minimum flow requirement for the riverine ecosystem was 100.13 m^3/s in the pre-dam period, although it was 70 m^3/s in the post-dam period. Due to flow regulation by the DVC dams, the minimum flow requirement for the aquatic habitat is also reduced because of decreased MAF. A threshold value for the riverine ecosystem, i.e., 10% of MAF, was not maintained in the lean months (December to May) in the pre-dam; however, it is maintained in the post-dam period based on the monthly average data at the Rhondia site of the Damodar River. Further, the environmental flow has been assessed after the downstream of major DVC dam sites and it is found that the optimum range of MAF is 259.2–432, 333–555, 1620–2700, and 2723.4–4539 MCM for Tilaiya, Konar, Maithon, and Panchet, respectively. Moreover, the poor or minimum flow (10% of MAF) is found to be 43.2, 55.5, 270, and 453.9 MCM for Tilaiya, Konar, Maithon, and Panchet, respectively, which are the minimum flows needed for sustaining or maintaining the aquatic habitat. The description of flow details is represented in Table 3.3.

Furthermore, the environmental flow has been assessed for different sites of the Damodar River from 1981 to 2010. It is found that the optimum range of MAF of five large dams was 41.89–69.83, 8.18–13.63, 6.22–10.37, 48.9–81.5, and 81.52–135.87 cumec for Tenughat, Konar, Tilaiya, Maithon, and Panchet dam sites, respectively, during 1981–2010. Besides, it was 19.3–32.16, 52.54–87.56, and 100.82–168.03 cumec for Barkisuriya, Phusro, and Damodar Bridge near Burnpur sites, respectively, during this period. Moreover, the poor or minimum flow was found to be 6.98, 1.36, 1.04, 8.15, and 16.8 cumec for five large dams of Tenughat, Konar, Tilaiya, Maithon, and Panchet, respectively and it was 3.22, 8.76, and 16.8 cumec for Barkisuriya, Phusro and Damodar Bridge near Burnpur sites, correspondingly, during this period, which is the minimum flow needed for sustaining or maintaining the aquatic habitat. The description of flow details is shown in Table 3.4.

Table 3.2 Environmental flow assessment at the Rhondia site of the lower Damodar River in the pre- and post-dam period

Months	Pre-dam (1934–1957)				Post-dam (1958–2007)			
	Monthly discharge (m^3/s)	MAF (m^3/s)	Optimum range (60–100%)	Minimum flow (10%)	Monthly discharge (m^3/s)	MAF (m^3/s)	Optimum range (60–100%)	Minimum flow (10%)
(Mar-May)	57.05	1001	600.80–1001.33	100.13	103.83	700.04	420.02–700.04	70
(Jun–Sept)	3371.2				2145.86			
(Oct–Nov)	477.05				433.48			
(Dec–Feb)	100.05				117			

Computed based on Bhattacharyya (2011) (Note: The mean annual flow (MAF) of the lower Damodar River at the Rhondia site (a threshold of 10% MAF is a prerequisite for the aquatic ecosystem)

Table 3.3 Assessment of environmental flow after the major DVC dams

Dam sites	Mean annual flow (MCM)	Periods	Description of flows							
			Flushing or maximum	Optimum range	Outstanding	Excellent	Good	Fair or degrading	Poor or minimum	Severe degradation
Tilaiya	432	October–March	864	259.2–432	172.8	129.6	86.4	43.2	43.2	<43.2
		April–September			259.2	216	172.8	129.6		
Konar	555	October–March	1110	333–555	222	166.5	111	55.5	55.5	<55.5
		April–September			333	277.5	222	166.5		
Maithon	2700	October–March	5400	1620–2700	1080	810	540	270	270	<270
		April–September			1620	1350	1080	810		
Panchet	4539	October–March	9078	2723.4–4539	1815.6	1361.7	907.8	453.9	453.9	<453.9
		April–September			2723.4	2269.5	1815.6	1361.7		

Computed based on Damodar Valley Corporation (2021)

Table 3.4 Environmental flow assessment in different sites from 1981 to 2010

Sites	MAF (cumec)	Periods	Description of flows							
			Flushing or maximum	Optimum range	Outstanding	Excellent	Good	Fair or grading	Poor or minimum	Serve degradation
Tenughat dam	69.83	October–March	139.66	41.89–69.83	27.93	20.95	13.96	6.98	6.98	<6.98
		April–September			41.89	34.92	27.93	20.95		
Konar dam	13.63	October–March	27.26	8.18–13.63	5.45	4.09	2.07	1.36	1.36	<1.36
		April–September			8.18	6.82	5.45	4.09		
Tilaya dam	10.37	October–March	20.79	6.22–10.37	4.15	3.11	2.08	1.04	1.04	<1.04
		April–September			6.22	5.19	4.15	3.11		
Barkisuriya	32.16	October–March	64.32	19.3–32.16	12.86	9.65	6.43	3.22	3.22	<3.22
		April–September			19.3	16.08	12.84	9.65		
Phusro	87.56	October–March	175.12	52.54–87.56	35.024	26.27	17.51	8.76	8.76	<8.76
		April–September			52.54	43.78	35.02	26.27		
Maithon dam	81.5	October–March	163	48.9–81.5	32.6	24.45	16.3	8.15	8.15	<8.15
		April–September			48.9	40.75	32.6	24.45		

(continued)

Table 3.4 (continued)

Sites	MAF (cumec)	Periods	Description of flows							
			Flushing or maximum	Optimum range	Outstanding	Excellent	Good	Fair or grading	Poor or minimum	Serve degradation
Panchet dam	135.87	October–March	271.74	81.52–135.87	54.35	40.76	27.17	13.59	13.59	<13.59
		April–September			81.52	67.94	54.35	40.76		
Damodar Bridge, Burnpur	168.03	October–March	336.06	100.82–168.03	67.21	50.41	33.61	16.8	16.8	<16.80
		April–September			100.82	84.02	67.21	50.41		

Computed based on Verma et al. (2015)

3.3 Role of Climate Change and LULC Dynamics on Flow Alterations

Climate change influences streamflow generation, basin hydrological processes, hydrological connectivity, and its control over ecosystem dynamics, as well as interactions between surface water and groundwater and the physical factors affecting streamflow disconnection (Islam et al., 2023; De Girolamo et al., 2022). It plays a significant role in altering environmental flows (Tang et al., 2015; Wang et al., 2023). Climate change impacts precipitation patterns, temperatures, and snowmelt, which in turn affect water availability and ecosystem health. These alterations lead to changes in precipitation patterns, resulting in increased flow variability and seasonal shifts in flow, which impact flow timing and volume (Pradhan et al., 2023). Moreover, rising temperatures increase evaporation, leading to greater water loss from water bodies and reduced flows. This temperature rise also contributes to thermal stress in water systems as warmer water holds less oxygen, posing risks to aquatic life, especially under low-flow conditions. Additionally, climate change influences the melting of glaciers and snow and contributes to sea-level rise. Sea-level rise causes saltwater intrusion into coastal rivers, while rising seas push saltwater further upstream, altering freshwater flow dynamics and affecting ecosystems (Hoque et al., 2025b). Moreover, climate change increases the frequency of extreme events, such as floods and droughts, which are common occurrences in the Damodar River Basin area.

The Land Use and Land Cover (LULC) dynamics play a crucial role in altering environmental flows in river basins. Changes in LULC, such as deforestation, agricultural expansion, urbanization, and wetland loss, significantly influence environmental flow, namely the quantity, quality, and timing of water flow in both surface and subsurface systems (Hoque et al., 2022b, 2024). Furthermore, it affects the hydrological cycle by impacting surface runoff, infiltration, evaporation, groundwater recharge, and flow patterns. Changes in LULC, particularly urbanization and construction of impervious surfaces, reduce the land's water absorption capacity, resulting in higher surface runoff, increased flood peaks, and shorter intervals between rainfall and runoff. Converting permeable land (forests, grasslands) to impervious surfaces limits water infiltration, reducing groundwater recharge and the baseflow contribution to streams, especially during dry periods. Changes in LULC, especially deforestation or conversion to built-up areas, reduce evapotranspiration, disrupting the water balance. Moreover, LULC changes lead to altered flow regimes, which include shifts in peak discharge (often increased), low flows (often decreased), seasonal flow patterns, flashiness, and flow variability. Deforestation and poor agricultural practices in the watershed increase soil erosion, leading to higher sediment loads in water bodies, which alter channel morphology and reduce flow capacity. Additionally, agricultural runoff, urban stormwater, and reduced vegetative buffers contribute to nutrient and pollutant loading (Hoque et al., 2023; Hoque & Islam, 2024). Hence, LULC dynamics have direct and profound effects on hydrological processes.

3.4 Improvement of Quantitative Environmental Flow

The DVC's dam construction, land-use changes (especially built-up areas), and climate change, combined with excessive water extraction, have significantly altered the quantitative flow regimes of the Damodar River (Mitra & Singh, 2018; Hoque et al., 2022a). These changes threaten the environment, riverine ecology, and biota of the river systems (Mitra & Singh, 2018; Hoque et al., 2022a; Purnanjali & Kasiviswanathan, 2022). Therefore, maintaining environmental flows is crucial for the sustainability of rivers and their ecosystem management. The DVC dams and barrages have notably altered the flow duration of the river, flood magnitude and frequency, which lead to more frequent low-flow events and fewer high-flow events after construction (Hoque et al., 2022a). Additionally, they have caused alterations in river morphology, like valley widening and rising sinuosity downstream, and landform development upstream, which denotes a fundamental change in the natural processes of the river (Ghosh & Guchhait, 2014). Large-scale water extractions and reservoir management are key human activities that degrade the hydrological health of the river and modify flow patterns (Mezger et al., 2021; Hoque et al., 2022a).

Future studies should be conducted on bias-corrected climate data for precipitation and air temperature to simulate climate scenarios and forecast future streamflow changes under various climate change conditions (Mineda et al., 2023). A comprehensive assessment is needed to explore how dam operations and climate patterns collectively affect environmental flows and baseline flow dynamics (Mezger et al., 2021; Mineda et al., 2023). Rising temperatures have been connected to reductions in summer baseflow, while other factors such as basin lithology, snowmelt timing and volume are key factors in rising spring baseflow (Murray et al., 2023). Several studies indicate that antecedent moisture remains a significant climate forecaster for monthly baseflow trends, highlighting the interconnection of baseflow dynamics and climate (Murray et al., 2023). Changes in water quality and flow threaten fish community structure and diversity, underscoring the need to extend assessments beyond just hydrology (Hoque et al., 2022a, 2025a). Given the diverse effects and exceptional characteristics of the river basin, a policy framework is essential to promote sustainable development and sustain ecological flows to protect river ecosystems (Hoque et al., 2022a; Purnanjali & Kasiviswanathan, 2022).

Recent findings on how changes in rainfall and temperature affect river flow and baseflow in monsoon-dependent regions such as the Damodar Basin need to be integrated (Chen et al., 2022; Hoque et al., 2022a; Mineda et al., 2023; Murray et al., 2023). This should include studies that surpass quantitative flow data to address effects on riverine ecosystems, biodiversity, and overall river health (Mitra & Singh, 2018; Hoque et al., 2022a). Discuss how changing flow regimes affect the livelihoods of local communities, like fishers, and how these insights can inform policy decisions for sustainable river management (Hoque et al., 2022a; Purnanjali & Kasiviswanathan, 2022). It should incorporate a variety of ecological or environmental flow assessment methods, including those focusing on habitat simulation and aggregate approaches and also flow duration curve techniques (Mitra & Singh, 2018). Thus, the next chapter, i.e., Chap. 4, is concerned with qualitative environmental flow analysis.

References

Babu, K. L., & Kumara, B. H. (2009). Environmental flows in river basins: A case study of river Bhadra. *Current Science*, 475–479.

Bhattacharyya, K. (2011). *The Lower Damodar River, India: Understanding the human role in changing fluvial environment*. Springer.

Blaney, D. (2013). Environmental flows, political dams. *Journal of Political Science and Public Affairs, 1*(2), 1–7.

Caissie, D., & El-Jabi, N. (1995). Comparison and regionalization of hydrologically based instream flow techniques in Atlantic Canada. *Canadian Journal of Civil Engineering, 22*(2), 235–246.

Caissie, D., El-Jabi, N., & Hébert, C. (2007). Comparison of hydrologically based instream flow methods using a resampling technique. *Canadian Journal of Civil Engineering, 34*(1), 66–74.

Chen, A., & Wu, M. (2019). Managing for sustainability: The development of environmental flows implementation in China. *Water (Basel), 11*(3), 433. https://doi.org/10.3390/w11030433

Chen, Z., Liu, Z., Yin, L., & Zheng, W. (2022). Statistical analysis of regional air temperature characteristics before and after dam construction. *Urban Climate, 41*, 101085. https://doi.org/10.1016/j.uclim.2022.101085

Damodar Valley Corporation. (2021). *DVC Reservoir Overview*. www.dvc.gov.in/dvcwebsite_newl/dams-barrages/DamodarValleyCoporation

De Girolamo, A. M., Barca, E., Leone, M., & Porto, A. L. (2022). Impact of long-term climate change on flow regime in a Mediterranean basin. *Journal of Hydrology: Regional Studies, 41*, 101061. https://doi.org/10.1016/j.ejrh.2022.101061

Ghosh, S., & Guchhait, S. K. (2014). Hydrogeomorphic variability due to dam constructions and emerging problems: A case study of Damodar River, West Bengal, India. *Environment, Development and Sustainability, 16*, 769–796. https://doi.org/10.1007/s10668-013-9494-5

Ghosh, S., Hoque, M. M., Saha, U. D., & Islam, A. (2022). Assessment of dam-induced changes in ecogeomorphological behaviour and fluvial functionality in the Damodar River, West Bengal, India. *AQUA—Water Infrastructure, Ecosystems and Society, 71*(6), 722–750. https://doi.org/10.2166/aqua.2022.003

Grantham, T. E., Viers, J. H., & Moyle, P. B. (2014). Systematic screening of dams for environmental flow assessment and implementation. *Bioscience, 64*(11), 1006–1018.

Hazra, S., Bhadra, T., Ghosh, S., & Barman, B. C. (2015). Assessing environmental flows for Indian sundarban: A suggested approach. *River Behaviour and Control, 35*, 65–74.

Hoque, M. M., Islam, A., & Ghosh, S. (2022a). Environmental flow in the context of dams and development with special reference to the Damodar Valley Project, India: A review. *Sustainable Water Resources Management, 8*(3), 1–27. https://doi.org/10.1007/s40899-022-00646-9

Hoque, M. M., Islam, A., & Mahammad, S. (2022b). Assessing the surface and bottom river water quality for drinking purpose and human health risk analysis: a study of Damodar River, India. *Arabian Journal of Geosciences, 15*(23), 1734. https://doi.org/10.1007/s12517-022-10987-6

Hoque, M. M., Islam, A., Islam, A. R. M. T., Pal, S. C., Mahammad, S., & Alam, E. (2023). Assessment of soil heavy metal pollution and associated ecological risk of agriculture dominated mid-channel bars in a subtropical river basin. *Scientific Reports, 13*(1), 11104. https://doi.org/10.1038/s41598-023-38058-0

Hoque, M. M., & Islam, A. (2024). Spatio-temporal dynamics of ecological, bacteriological, and overall water quality of the Damodar River, India. *Environmental Science and Pollution Research, 31*(12), 18465–18484. https://doi.org/10.1007/s11356-024-32185-5

Hoque, M. M., Islam, A., Sarkar, B., & Saha, U. D. (2024). Assessing the surface and bottom river water quality for irrigation: A study of Damodar River, India. *International Journal of Energy and Water Resources, 8*(4), 545–562. https://doi.org/10.1007/s42108-022-00206-z

Hoque, M. M., Islam, A., Pal, S. C., Sarkar, B., & Hossain, Z. M. (2025a). Assessment of fish habitat suitability in the wake of elevated pollution of a tropical river in India. *Environmental Science and Pollution Research, 32*(3), 1208–1222. https://doi.org/10.1007/s11356-024-35685-6

Hoque, M. M., Pal, S. C., Mandal, S., & Islam, A. (2025b). Assessment of mollusc habitat suitability of an anthropogenically altered tropical river in India. *Environmental Monitoring and Assessment, 197*(10), 1–18. https://doi.org/10.1007/s10661-025-14593-3

Islam, A., Sardar, N., Mohinuddin, S., Hoque, M. M., Sengupta, S., Das, B. C., ... & Sengupta, B. (2023). Quasi-equilibrium channel metamorphosis in planform of a subtropical river in India in post-dam period. *Catena, 221*, 106793. https://doi.org/10.1016/j.catena.2022.106793

Jain, S. K. (2015). Assessment of environmental flow requirements for hydropower projects in India. *Current Science, 108*(10), 1815–1825.

Jain, S. K., & Kumar, P. (2014). Environmental flows in India: Towards sustainable water management. *Hydrological Sciences Journal, 59*(3–4), 751–769.

Jha, R., Sharma, K. D., & Singh, V. P. (2008). Critical appraisal of methods for the assessment of environmental flows and their application in two river systems of India. *KSCE Journal of Civil Engineering, 12*, 213–219.

Karim, K., Gubbels, M. E., & Goulter, I. C. (1995). Review of determination of instream flow requirements with special application to Australia. *Journal of the American Water Resources Association, 31*(6), 1063–1077. https://doi.org/10.1111/j.1752-1688.1995.tb03422.x

Karr, J. R. (1991). Biological integrity: A long-neglected aspect of water resource management. *Ecological Applications, 1*(1), 66–84.

Mezger, G., del Tánago, M. G., & De Stefano, L. (2021). Environmental flows and the mitigation of hydrological alteration downstream from dams: The Spanish case. *Journal of Hydrology, 598*, 125732. https://doi.org/10.1016/j.jhydrol.2020.125732

Mineda, H., Nukazawa, K., & Suzuki, Y. (2023). Assessing environmental flow alterations induced by dams and climate change using a distributed hydrological model at catchment scale. *Journal of Water and Climate Change, 14*(12), 4821–4836. https://doi.org/10.2166/wcc.2023.538

Mitra, S., & Singh, A. (2018). Assessment of environmental flow requirements of Damodar river basins by using flow duration indices method—A case study. *International Journal of Hydrology, 2*(3), 281–283. https://doi.org/10.15406/ijh.2018.02.00081

Murray, J., Ayers, J., & Brookfield, A. (2023). The impact of climate change on monthly baseflow trends across Canada. *Journal of Hydrology, 618*, 129254. https://doi.org/10.1016/j.jhydrol.2023.129254

Pandey, R. P., & Ramasastri, K. S. (2003). Estimation of lean season water availability in streams with limited data. *Journal of the Institution of Engineers. India. Civil Engineering Division, 84*, 149–152.

Poff, N. L., Richter, B. D., Arthington, A. H., Bunn, S. E., Naiman, R. J., Kendy, E., et al. (2009). The ecological limits of hydrologic alteration (ELOHA): A new framework for developing regional environmental flow standards. *Freshwater Biology, 55*(1), 147–170. https://doi.org/10.1111/j.1365-2427.2009.02204.x

Pradhan, P., Shrestha, S., & Shanmugam, M. S. (2023). Potential impacts of climate change on ecologically relevant flows: A case study in the Himalayan River basin. *AQUA—Water Infrastructure, Ecosystems and Society, 72*(11), 2109–2125. https://doi.org/10.2166/aqua.2023.284

Purnanjali, C., & Kasiviswanathan, K. S. (2022). Assessment of impacts of altered environmental flow on fishing in lower Damodar river basin, India. *Acta Geophysica, 70*(2), 833–846. https://doi.org/10.1007/s11600-022-00752-4

Reiser, D. W., Wesche, T. A., & Estes, C. (1989). Status of instream flow legislation and practices in North America. *Fisheries, 14*(2), 22–29.

Sarkar, B., & Islam, A. (2022). Assessing poverty and livelihood vulnerability of the fishing communities in the context of pollution of the Churni River, India. *Environmental Science And Pollution Research, 29*(18), 26575–26598. https://doi.org/10.1007/s11356-021-17719-5

Shokoohi, A., & Hong, Y. (2011). Using hydrologic and hydraulically derived geometric parameters of perennial rivers to determine minimum water requirements of ecological habitats (case study: Mazandaran Sea Basin—Iran). *Hydrological Processes, 25*(22), 3490–3498.

Silk, N., & Ciruna, K. (2013). *A practitioner's guide to freshwater biodiversity conservation*. Island Press.

Smakhtin, V., & Anputhas, M. (2006). *An assessment of environmental flow requirements of Indian river basins* (Vol. 107). International Water Management Institute.

Smakhtin, V. U., & Eriyagama, N. (2008). Developing a software package for global desktop assessment of environmental flows. *Environmental Modelling & Software, 23*(12), 1396–1406.

Smakhtin, V., Revenga, C., & Döll, P. (2004). A pilot global assessment of environmental water requirements and scarcity. *Water International, 29*(3), 307–317.

Smakhtin, V., Arunachalam, M., Behera, S., Chatterjee, A., Das, S., Gautam, P., et al. (2007). *Developing procedures for assessment of ecological status of Indian river basins in the context of environmental water requirements* (Vol. 114). IWMI.

Tang, J., Yin, X. A., Yang, P., & Yang, Z. F. (2015). Climate-induced flow regime alterations and their implications for the Lancang river, China. *River Research and Applications, 31*(4), 422–432. https://doi.org/10.1002/rra.2819

Tegos, A., Schlüter, W., Gibbons, N., Katselis, Y., & Efstratiadis, A. (2018). Assessment of environmental flows from complexity to parsimony—Lessons from Lesotho. *Water, 10*(10), 1–17. https://doi.org/10.3390/w10101293

Tharme, R. E. (2003). A global perspective on environmental flow assessment: Emerging trends in the development and application of environmental flow methodologies for rivers. *River Research and Applications, 19*(5–6), 397–441.

The Brisbane Declaration. (2007). Environmental flows are essential for freshwater ecosystem health and human well-being. In *The 10th International Riversymposium and International Environmental Flows Conference, Brisbane, Australia* (pp. 3–6).

Verma, R. K., Murthy, S., & Tiwary, R. K. (2015). Assessment of environmental flows for various sub-watersheds of damodar river basin using different hydrological methods. *International Journal of Waste Resources, 5*(4), 1–6.

Wang, Q., Liu, Y., Wang, Y., Zhang, Y., Li, L., & Wang, L. (2023). Separating climate change and human activities' effects on flow regime with hydrological model error correction. *Ecological Indicators, 157*, 111265. https://doi.org/10.1016/j.ecolind.2023.111265

Yulianti, J. S., & Burn, D. H. (1998). Investigating links between climatic warming and low streamflow in the Prairies region of Canada. *Canadian Water Resources Journal, 23*(1), 45–60.

Chapter 4
Damodar Valley Corporation and Qualitative Environmental Flow Alteration

4.1 Dams and Qualitative Flow Alterations in the Global and South-East Asian Context

Environmental flow emphasizes the quantitative and qualitative dimensions of water flow required for aquatic ecosystems and many developmental activities such as agriculture, industry, energy, etc (Hoque et al., 2022a). The anthropogenic interventions in terms of pollution and flow modification alter the river's environmental flows, which negatively affect the riverine ecosystems and biodiversity and the livelihood of people who depend on the healthy ecosystems (Ghosh et al., 2022; Hoque et al., 2022a). Pollution from various points and nonpoint sources such as urban-industrial effluents, overexploitation of water (Falowo et al., 2017), mining activities (Singh et al., 2014), and agrochemicals from agricultural fields (Hoque et al., 2023a; Al-Hadithi et al., 2019) causes polluted river water quality. The polluted river impacts agricultural production, sustainability, and development by exploiting crop quality, soil health (Sarkar & Islam, 2019), and also human health. Moreover, drinking water quality assessment is essentially required for third-world countries, especially in India, as many people are suffering from water-related diseases (Kumar et al., 2022). Besides, the alteration of environmental flow impacts the riverine ecology and biota. Thus, it is required to assess the river water quality from the perspective of irrigation water quality for sustainable agricultural development, drinking water quality for human health risks and ecological water quality for riverine ecology and biota. Therefore, the water quality of the Damodar River is examined for addressing three broad areas: (a) evaluating irrigation water suitability, (b) drinking water suitability with human health risk analysis, and (c) ecological water quality assessment to give a broader picture of the deterioration of water quality in terms of qualitative environmental flow perspective.

Supplementary Information The online version contains supplementary material available at https://doi.org/10.1007/978-3-032-11161-6_4.

M. M. Hoque, A. Islam, *Dams, Development and Environmental Flow in the Damodar River Basin*, https://doi.org/10.1007/978-3-032-11161-6_4

4.2 Alteration of Irrigation Water Quality

Pollution of river water is a global environmental issue (Islam et al., 2024). The quality of river water for different uses like industrial uses, irrigation, aquatic habitat, and human activities such as swimming, bathing, and boating, can be determined through a physiochemical examination of the status of the water and it is vital since it indicates the extent of water pollution (Sarkar & Islam, 2019). The extensive use of river water for irrigation harms agricultural production, the exploitation of water quality and its sustainability by decreasing crop quality and soil health, along with the agricultural outcomes (Sarkar & Islam, 2019). Furthermore, not only the quality of water delivered but also the quantity of water given, as well as a better irrigation system, are all compulsory for sustainable agricultural development (Sarkar et al., 2022). Hence, there is a global need to create an irrigation system for sustainable agriculture enclosed by a watershed (Spandana et al., 2013).

The river water quality declines as a consequence of mining activities, industrial activities, agricultural activities and urbanogenic effluents, and the use of surface water enclosed by a watershed, which brings about overexploitation of river water (Al-Hadithi et al., 2019; Singh et al., 2014). Hence, the river water deterioration, both quantitative and qualitative and its impact are viewed from the channel and watershed scale. The aquatic ecosystems are thus threatened due to the deteriorating river water quality (Deepika & Singh, 2015). At the watershed level, it impacts the quality and productivity of crops because of poor irrigation water quality (Chen et al., 2003). Researchers have thus shown huge concern for evaluating the suitability of river water for irrigation. There are several indices related to permeability, alkalinity, sodicity, magnesium hazard, and salinity hazard which are widely used for assesing the irrigation hazards at the global scale (Eaton, 1950; Yıldız & Karakuş, 2020; Raghunath, 1987; Kelley, 1963; Todd, 1960; Sarkar & Islam, 2019; Dhembare, 2012; Tiri et al., 2020). However, the comprehensive irrigation water quality assessment at the basin scale of the Damodar River is not attempted systematically.

Therefore, the primary research objectives set for this study are: (1) To figure out the irrigation hazards of the surface and bottom water for assessing irrigation water suitability and (2) To identify the vertical and spatial variation in irrigation suitability and water quality based on data collected in February 2020 from the bottom and the surface of the Damodar River and its tributaries. At each of the 14 selected stations, a total of 28 water samples were taken from the bottom (14 samples) and surface (14 samples) of the rivers. The Damodar River water quality for irrigation has been determined using various kinds of irrigation hazards (like specific toxicity, salinity, alkalinity, and sodicity) and a irrigation water quality index (IWQI).

4.2.1 Irrigation Hazards

Based on the status of the Sodicity hazard ((a) SAR (b) %Na (c) SSP (d) KR), Alkalinity hazard ((a) RSC (b) RSBC), Permeability Index (PI), Salinity hazard ((a) EC and TDS (b) PS), and Magnesium hazard (MH) or Magnesium absorption ratio (MAR), the irrigation hazard has been calculated in the study area.

4.2.1.1 Sodicity Hazard

1. **Sodium adsorption ratio (SAR):** The SAR is an indicator of sodium hazard, which impacts crops and soil permeability. The total salinity of the water and the soil permeability are impacted by the higher percentage of sodium in irrigation water. The river surface water sample's SAR values range from 0.38 to 1.12 with a mean value of 0.66, whereas the river bottom water sample's range from 0.40 to 1.04 with a mean value of 0.67 (Fig. 4.2h). According to the SAR values, irrigation water is classified as (SAR > 9) unsuitable, (6 < SAR < 9) doubtful, and (SAR < 6) good (Yıldız & Karakuş, 2020). Every sample for surface and also river bottom water is suitable for irrigation as they are under good categories (Table 4.S1). Moreover, using the U.S Salinity Laboratory (USSL) diagram plotted against the EC with the help of SAR, the suitability of river water for irrigation has been established. According to the diagram, it was determined that all the samples (both the surface and bottom water) fall within S1, i.e., low sodium hazard, which indicates that all the sample locations are not influenced by sodium hazard (Fig. 4.2a, b).
2. **Percentage of sodium (%Na):** The %Na of surface water samples ranged from 14.87 to 35.08, with a mean value of 23.83 and for the bottom water samples ranged from 15.85 to 33.20, with a mean value of 24.64 (Fig. 4.2f). Based on the classification of %Na for the irrigation water suitability, 35.71% ($N = 5$) of the samples show excellent, and 64.29% ($N = 9$) show good for river surface water. In addition, 28.57% ($N = 4$) is excellent and 71.43% ($N = 10$) is good regarding river bottom water samples (Table 4.S1). Hence, the %Na of the surface and river bottom water samples shows that most of the samples are suitable for irrigation. Additionally, a Wilcox diagram is created to graphically appraise the river water suitability for irrigation, where the %Na was plotted against the concentration of EC. It is found that every sample falls under the excellent to good category (Fig. 4.1a, b).
3. **Soluble sodium percentage (SSP):** The SSP has values between 14.36% and 49.29% with a mean value of 26.59% of the surface water samples, whereas river bottom water samples have SSP values between 15.33% and 43.93% with a mean value of 27.52% (Fig. 4.2l). If the SSP is greater than 50%, it indicates that the river water is unsuitable for irrigation and harmful; if it is less than 50%, it indicates irrigation suitability. Out of the total samples with SSP values measured for the surface waters, 4 (28.57%) are in the excellent category, while 9 (64.29%) are in the good category, and 1 (7.14%) was found permissible for irrigation. On the contrary, for samples of the river bottom water, 3 (21.43%) are in the excellent category, whereas 9 (64.29%) are in the good category, and 2 (14.29%) are found permissible for irrigation (Table 4.S1). Moreover, the SSP metrics illustrate that every sample taken from the river's surface and bottom at the 14 selected stations for the study is fit for irrigation.
4. **Kelly ratio (KR):** With the help of the KR, the river water suitability for irrigation is also estimated. For surface water, the calculated KR values range between 0.15 and 0.51 with a mean value of 0.28 and between 0.16 and 0.45 with a mean value of 0.29 in the river bottom water (Fig. 4.2k). The KR concentrations in the river's surface and bottom water reveal the narrowest difference. According to

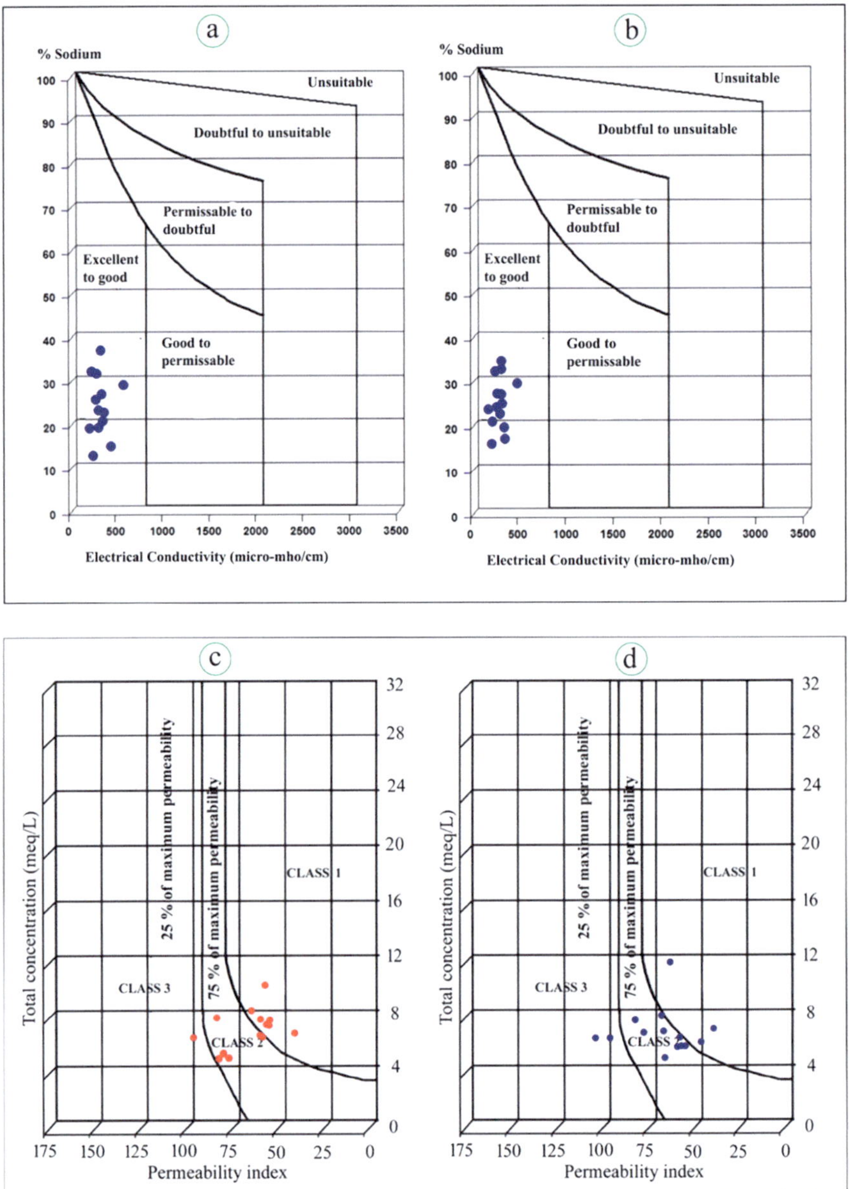

Fig. 4.1 Suitability of the Damodar River and its tributaries' water for irrigation; (**a**) Wilcox diagram (surface water), (**b**) Wilcox diagram (bottom water), (**c**) Doneen's permeability index (surface water), (**d**) Doneen's permeability index (bottom water)

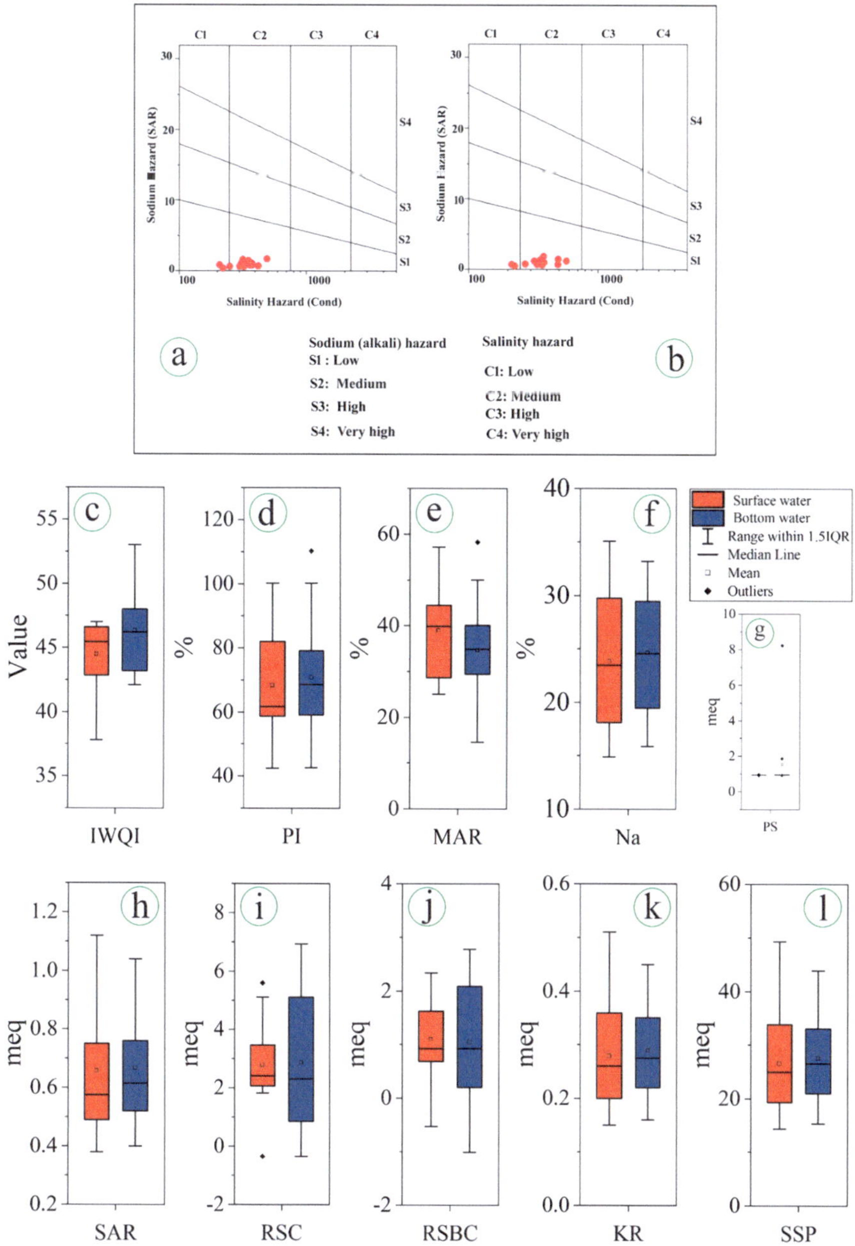

Fig. 4.2 United States Salinity Laboratory (USSL) classification; (**a**) surface water, (**b**) bottom water, and (**c–l**) represent the categorical variation of the irrigation water quality index (IWQI) and different indices of surface water and bottom water samples

(Dhembare, 2012), river water that has a KR greater than 1 is unsuitable and harmful for irrigation, whereas, KR less than 1 indicates the reciprocal. A surplus concentration of Na^+ in water is indicated even if the KR is greater than 1 (Tiri et al., 2020). The calculated KR values reveal that every water sample taken for irrigation is suitable (Table 4.S1) and there is no excess Na^+ concentration in the river water of the sites under study.

4.2.1.2 Alkalinity Hazard

Alkalinity hazard impacting the growth of the crops is measured by the indices of RSC and RSBC discussed in the following sections.

1. **Residual sodium carbonate (RSC):** The concentration of bicarbonate, carbonate, sodium, and magnesium in the river water is measured to determine the alkalinity hazard. The values of the surface river water ranged between −0.35 and 5.60 meq/L, with a mean value of 2.78 meq/L for RSC. Furthermore, the bottom water samples ranged between −0.35 and 6.94 meq/L with a mean value of 2.87 meq/L (Fig. 4.2i).

   According to the RSC values, RSC < 1.25 meq/L indicates good, 1.25 < RSC < 2.5 meq/L indicates medium, and RSC > 2.5 meq/L indicates bad, for the suitability of water for irrigation (Table 2). 1 (7.14%) indicates good, 6 (42.86%) indicates medium, and 7 (50%) indicates bad water condition in terms of irrigation suitability among the 14 surface water samples. Moreover, 6 (42.86%) indicates good, 1 (7.14%) indicates medium, and 7 (50%) indicates bad water condition in terms of irrigation suitability among the 14 bottom water samples (Table 4.S1).
2. **Residual sodium bicarbonate (RSBC):** The RSBC is also used to evaluate the alkalinity hazards but based on calcium and bicarbonate concentration. According to the measured RSBC, the RSBC of river surface water ranges between −0.53 and 2.34 meq/L with a mean value of 1.10 meq/L and RSBC of bottom water ranges between −1.01 and 2.78 meq/L with a mean value of 1.04 meq/L (Fig. 4.2j).
3. **Permeability index (PI):** The PI is a significant index used for calculating the irrigation water's suitability (Batarseh et al., 2021). The surface water's PI values range between 42.45% and 100.27% with a mean value of 68.34%, and similarly, the bottom water's PI values range between 42.54% and 110.28% with a mean value of 70.86% (Fig. 4.2d). The mean value of PI of surface water samples is lower than the bottom water samples. According to the threshold values of PI, 9 (64.29%) water samples are suitable, and 5 (35.71%) samples are in good condition for irrigation among the 14 surface water samples for the status of irrigation suitability. Moreover, 10 (71.43%) samples can be estimated as suitable and 4 (28.57%) for irrigation for the river bottom water, which were found in good condition (Table 4.S1).

Using Doneen's diagram, the PI values and the salt concentration tested, 28 samples have been plotted. It shows that only 1 water sample (7.14%) and 2 samples (14.29%) of the surface water and the river bottom, respectively, fall under class 3 (i.e., 25% of maximum permeability), which means they are not suitable for irrigation. Besides, 6 (42.86%) of the surface water and 9 samples (64.29%) of river bottom, fall under the class 2 category (i.e., 75% of maximum permeability), which means they are suitable for irrigation, whereas, 7 samples (50%) and 3 (21.43%) of the surface and the river bottom respectively, fall under class 1 (i.e., more than 75% of maximum permeability), that means they are good water for irrigation (Fig. 4.1c, d). Therefore, compared to the bottom water of the river, the surface water has shown a higher degree of suitability for irrigation.

4.2.1.3 Salinity Hazard

1. **Electrical conductivity (EC) and total dissolved solids (TDS) analysis:** The EC and TDS concentrations in water can be estimated as the salinity threat (Sarkar & Islam, 2019). It is not suitable to use a higher concentration of EC for irrigation. The EC ranges between 230 and 519 μS/cm with a mean value of 328.71 μS/cm for the surface water samples, and the EC ranges between 225 and 570 μS/cm with a mean value of 378.789 μS/cm for the bottom water samples. Hence, in comparison to the surface water of EC, the mean value of the bottom water is higher. Based on the suitability of water for irrigation according to EC (Table 4.S1), 14.29% of the total samples indicate excellent and 85.71% indicate good revealed by the surface water samples for irrigation. Moreover, 7.14% are determined excellent of the total samples related to the bottom water samples, and 92.86% indicate good for irrigation.

   In irrigation water, high TDS concentration influences saline hazards (Tiri et al., 2020). The TDS varies between 161 and 293 mg/L with a mean value of 210.57 mg/L, while it ranges between 151 and 490 mg/L with a mean value of 223.14 mg/L in the surface water samples and the bottom water samples, respectively. Compared to surface water samples, river bottom water samples have a greater TDS concentration. Moreover, all of the surface water samples are formed as excellent for irrigation, as per the TDS level (Table 4.S1). Apart from this, 92.86% of the bottom samples are excellent for irrigation, and the remaining 7.14% are recognized as good for irrigation.

   According to Richards (1954), the EC, the concentration of soluble salt, is classified into four categories for calculating the sodium hazard and salinity of irrigation water. Very high or C4 (EC >2250 μS/cm), high or C3 (EC 750–2250 μS/cm), medium or C2 (EC 250–750 μS/cm), and low or C1 (EC <250 μS/cm) have been categorized for salinity hazards. Similarly, very high or S4 (SAR >26), high or S3 (SAR 18–26), medium or S2 (SAR 10–18), and low or S1 (SAR <10) for sodium hazards. In the U.S. Salinity Laboratory Staff (USSL) diagram, all the

EC values were plotted against the SAR value. Between the C1 to C2 and the S1 class, all the tested water samples found which show low to medium salinity hazards (Fig. 4.2a, b). Furthermore, not only the surface water but also the bottom water records 14.29% of samples that show low salinity hazard (C1). Moreover, both the surface and bottom water records as 85.71% of samples, which shows a medium salinity hazard (C2). The Barakar River shows a low level of salinity hazard (C1) (an average EC of 240.5 μS/cm) as per portrayed on the USSL diagram, whereas, a medium level of salinity hazard (C2) (an average EC of 544.5 μS/cm) portrayed for the Garga River.

2. **Potential salinity (PS):** Irrigation water suitability is also measured with the help of PS. PS ranges between 0.92 and 0.96 meq/L with a mean value of 0.95 meq/L for the surface water samples, whereas it ranges between 0.93 and 8.23 meq/L with a mean value of 1.54 meq/L for the bottom water samples. The mean concentration shows bottom river water is much higher than surface water samples of potential salinity (Fig. 4.2g). The tested water sample's PS levels are classified depending on their degree of suitability for irrigation (Table 4.S1). All the samples are either excellent or good for irrigation; only one bottom water sample was determined as unsatisfactory for irrigation. Therefore, the overall comparison suggests that the surface water for irrigation has better suitability than the bottom water.

4.2.1.4 Magnesium Hazard

The soil structure is damaged due to the high concentration of magnesium in irrigation water. Because of the high water absorption between magnesium and clay particles, a high concentration of magnesium in irrigation water results in a decline in the infiltration capacity of soil and soil alkalinity. The water is safe for irrigation as the value indicates MH <50%. The MH ranges between 25.06% and 57.22% with a mean value of 38.94% for surface water and between 14.53% and 58.30% with a mean value of 34.65% for river bottom water (Fig. 4.2e). Hence, the surface water samples show a greater concentration of MH than the bottom water samples. Furthermore, in both the surface and bottom water of the river, the MH values of 12 (85.71%) samples fall under the permissible condition (<50%) (Table 4.S1). Hence, there is no difference in the suitability of water for irrigation even if the surface water has a greater MH concentration than the bottom water samples.

4.2.2 *Irrigation Water Quality Index*

One of the effective tools used by policymakers to analyse irrigation water quality is the IWQI (Batarseh et al., 2021). Even the degree of river water suitability for irrigation is also measured with the help of IWQI. In the present work, the IWQI of all the water samples is unsuitable to very poor as it ranges between 37.83 and

53.02. Furthermore, the IWQI is unsuitable for very poor water as it ranges between 37.83 and 47.01 with a mean value of 44.50 for surface river water, whereas the IWQI is very poor as it ranges between 42.10 and 53.02 with a mean value of 46.31 for bottom river water (Fig. 4.2c). Both the water samples are unsuitable to very poor for irrigation in terms of IWQI as per the irrigation water quality standard. All the river bottom water samples are very poor for irrigation, whereas among the 14 surface water samples, 85.71% (12 samples) are determined as very poor for irrigation. On the other hand, the rest are found in the unsuitable category for irrigation. (Table 4.S1).

4.3 Alteration of Drinking Water Quality and Perspectives of Human Health

One of the most important resources is water, as all living beings depend on it for their existence on the earth (Eslamian et al., 2018). Food processing, bathing, drinking, and irrigation are human activities that use water (Meinzen-Dick & Bakker, 2001). Therefore, for a healthy environment and human body, water quality needs to be examined (Hoque et al., 2023b). Especially in India, the assessment of water quality for drinking purposes is highly required because most of the people suffer from waterborne diseases, as India belongs to third-world countries (Kumar et al., 2022). Therefore, water quality assessment is essential due to human health concerns (Sarkar & Islam, 2019; Eslamian et al., 2018). Most of the people of rural areas in developing countries, such as India, rely on either surface water or groundwater, like ponds, rivers, and lakes (Sarkar & Islam, 2021). High concentrations of pollutants in drinking water may cause several diseases like heart diseases, blue-baby syndrome, kidney stones, teeth, bones, methemoglobinemia, hypertension, gastric intestinal cancer, catharsis, vomiting, Blackfoot disease and other skin diseases, diarrhoea, birth defects, high blood pressure, and even cancer (Sunitha et al., 2021). Hence, water quality and human health are linked directly (Sunitha et al., 2021; Ostad-Ali-Askari et al., 2017). Over the past two decades (2000–2020), fluoride and nitrate-induced water contamination have become a serious human health concern worldwide (Duvva et al., 2022; Adimalla, 2020). High concentration of fluoride and nitrate in drinking water consumption results in a negative impact on human health and have recently become a concerning issue (Duvva et al., 2022). Natural processes and human activities result in higher concentrations of fluoride and nitrate in river water.

As it is a significant supply of drinking water, the quality of the water must be examined to monitor human health (Ahmed et al., 2020). It is also necessary to evaluate controlling the pollution of water and the level of risk of getting safe drinking water. Thus, examining river water quality is essential for raising public health awareness and the remediation of river water resources for sustainable management (Chakraborty et al., 2021a, 2021b; Ostad-Ali-Askari & Shayannejad, 2021). In this regard, numerous researchers' work not only on groundwater but also on surface water for drinking purposes has been conducted (e.g., Mahammad & Islam, 2021; Sarkar & Islam, 2021). In recent times, in the sub-tropical and tropical regions of the Earth, numerous researchers have calculated the water quality based on the concentration of nitrate and fluoride (Jannat et al., 2022; Bera et al., 2021). Furthermore, to evaluate the amount of dissolved oxygen (DO), the L-moment method has been applied, which is crucial for water quality assessment with regard to the river ecosystems in the Sakarya Basin, Turkey (Citakoglu et al., 2021). Likewise, the DO sag curve providing valuable insights on the exploitation of water quality in relation to drinking purposes and biological requirements is examined for Churni River in India (Sarkar et al., 2021). However, there is a limited understanding on the drinking water quality assessment in the context of the DRB.

Thus, the objectives of the present research would systematically address: (1) To demonstrate the quality of the bottom and surface water of Damodar River with the help of physico-chemical properties and ionic chemistry; (2) To evaluate the river water suitability for drinking and health-related risks, including carcinogenic and non-carcinogenic impacts, using water quality indices based on the data collected from the Damodar River and its tributaries in February 2020. A total of 28 water samples were taken, 14 samples from the surface and the other 14 from the bottom of the river.

4.3.1 *Drinking Water Quality Analysis*

According to the chemical concentration of every parameter as specified by the standard permissible limits of BIS (2012), the drinking quality of both surface and bottom water has been examined in this study. Based on the BIS standard (2012), the water quality values are measured and given in Table 4.S3. In both surface and bottom water, the parameters that are out of their limits have also been displayed in Table 4.S2. The chemical concentration, in the surface water of EC (78.57%) and total hardness (TH) (14.29%), overcame their safe limits, whereas the concentrations of EC (78.57%), TH (14.29%), Cl^- (14.29%), and alkalinity (7.14%) in bottom water overcame their safe limits. Moreover, the chemical concentration of not only the surface but also the bottom water of each parameter was within safe limits.

In addition, in the water sample in the basin area, the concentration of EC (76.47%), pH (17.65%), Mg^{2+} (91.18%), Ca^{2+} (14.71%), K^+ (5.88%), and DO (2.94%) exceeded their safe limits. However, the water samples fall under the safe limits of the concentration of TH, TDS, total suspended solids (TSS), Na^+, and total alkalinity (TA).

The river water quality of the surface and the bottom water has been measured with the help of standard permissible limits of all the parameters using the weighted arithmetic water quality index (WQI) for drinking purposes (Brown et al., 1972). The WQI values are measured and related to the criteria of water quality, and according to their status, water quality values have been categorized (Table 4.S3). In this study, the WQI values were categorized into five classes as excellent (0–25), good (25–50), poor (50–75), very poor (75–100), and unsuitable ($\geq$ 100). The surface water quality ranges between 37.86 and 72.25, which shows good to poor, found in the present study, whereas the bottom water samples WQI ranges between 37.95 and 79.4, which shows good to very poor, in the present study. According to the mean measured values of water samples, in comparison, surface water is more suitable for drinking than bottom water. The bottom river water has 79.4 as the highest WQI value of the Jamunia River (sample location 11), and the status of water quality is found to be very poor. Likewise, in the Konar River, the surface water has the lowest WQI of 37.86, and the status of the water quality is good. According to the present work, 9 (64.29%) and 5 (35.71%) tested water samples relate to poor and good water quality status in surface water; whereas, in the bottom water, the water quality relates to 7 (50%) as good, 6 (42.86%) as poor, and 1 (7.14%) as very poor tested water samples. By comparing, the WQI values for drinking purposes show that the surface water quality is better than the bottom water. Furthermore, of all the 28 samples, a sum of 15 (53.57%) as good, 12 (42.86%) as poor and 1 (3.57%) was very poor tested water samples. Similarly, in this study's findings, Singh et al. (2019) reported that the upper and lower stretches of the Damodar River's water quality range between clean and contaminated, which is not fit for drinking purposes. Thus, at each sampling station, the water quality status of the river water is rated as nearly good to poor for drinking purposes. Therefore, in view of the increasing level of pollution, the river water's environment and its quality are deteriorating day by day.

4.3.2 Human Health Risk Analysis

High concentrations of pollutants in drinking water may cause several diseases like heart diseases, blue-baby syndrome, kidney stones, teeth, bones, methemoglobinemia, hypertension, gastric intestinal cancer, catharsis, vomiting, Blackfoot disease and other skin diseases, diarrhoea, birth defects, high blood pressure, and even cancer (Sunitha et al., 2021). For drinking and other household activities, in developing and underdeveloped countries majority of the people, specifically the rural population, rely on rivers and groundwater. Therefore, in rural populations, a major severe issue in water due to fluoride (F^-) and nitrate (NO_3^-) in water causes the non-carcinogenic risk. For instance, the south Indian population suffers from illness due to high concentrations of F^- and NO_3^- in drinking water (Sunitha et al., 2021). Thus, calculating the concentration of F^- and NO_3^- in drinking water is very crucial as it causes severe health risks.

4.3.2.1 Non-carcinogenic Risk Assessment

However, the physiochemical parameters and drinking water analysis have already been done in the last section; the present study has examined the non-carcinogenic parameters of NO_3^- and F^- for the non-carcinogenic risk of babies, children, and adults by the United States Environmental Protection Agency (USEPA). ~46% of water samples are categorized as poor water quality. However, the concentration of non-carcinogenic parameter values of NO_3^- and F^- belongs to their permissible limit. Therefore, not a single water sample belonged to non-carcinogenic health risks. Likewise, in the present study findings, Maleki and Jari (2021) evaluated human health risks (carcinogenic and non-carcinogenic) for the rural people of Iran, Kurdistan and they noted that the HQ value of all the metal for women, men and children is >1, i.e., permissible limits for drinking water and they also explained that there is only arsenic metal which causes health issues for water consumers in terms of the non-carcinogenic risk, no other issues has been found. However, non-carcinogenic human health risk has been examined to evaluate the level of health risk for surface as well as bottom water samples through the total hazard index (THI) and hazard quotient (HQ).

This study observed that HQ values of F^- in the surface water sample ranging between 0.13 and 0.45 (adults), 0.13 and 0.46 (children), and 0.10 and 0.35 (babies) with an average value of 0.25, 0.26, and 0.20, respectively, (Table 4.S4) whereas, HQ values in the bottom water sample ranging between 0.13 and 0.51 (adults), 0.13 and 0.52 (children), and 0.10 and 0.40 (babies) with an average value of 0.26, 0.26, and 0.20, respectively, (Table 4.S4). The HQ values of NO_3^- ranging between 0.04 and 0.10 (adults), 0.04 and 0.10 (children), and 0.03 and 0.08 (babies) with an average value of 0.07, 0.07, and 0.05, respectively, in the surface water sample (Table 4. S4), while, the HQ value ranging between 0.03 and 0.11 (adults), 0.03 and 0.11 (children), and 0.02 and 0.09 (babies) with an average value of 0.07, 0.07 and 0.05, respectively, in the bottom water samples, in the study area (Table 4.S5). Moreover, based on the total HQ values of F^- and NO_3^-, THI values have been calculated. In the surface water, the THI values range from 0.16 and 0.50 with an average value of 0.32 (adults), 0.17 and 0.51 with an average value of 0.32 (children), and 0.13 and 0.39 with an average value of 0.25 (babies) (Table 4.S4); whereas, THI values for the bottom water sample ranging between 0.16 and 0.62 with an average value of 0.33 (adults), 0.16 and 0.63 with an average value of 0.33 (children), and 0.12 and 0.48 with an average value of 0.25 (babies) (Table 4.S5). Therefore, the bottom water samples are higher than in the surface water, as per the values of HQ (both non-carcinogenic parameters of F^- and NO_3^-) and THI. The present study is free from non-carcinogenic health risks as both the values of HQ as well as THI come under their permissible limit (1), thus surface water is better than the bottom water of the river.

4.3.2.2 Carcinogenic Risk Assessment

In this work, the carcinogenic risk (CR) is calculated using CR values of NO_3^- for adults, children, and babies in both levels of water samples with the help of the recommended approach of USEPA. The carcinogenic risk from NO_3^- in river water has been assessed for adults, children, and babies (Table 2.S6). The present work found that the CR value for NO_3^- ranges from 5.769×10^{-7} to 1.6154×10^{-6} in the surface water samples, while it ranges from 5.000×10^{-7} to 1.8077×10^{-6} in the bottom water samples for adults. For children, it varies from 5.85×10^{-7} to 1.638×10^{-6} in the surface water samples, whereas 5.07×10^{-7} to 1.833×10^{-6} in the bottom water samples. For babies, the carcinogenicity average for NO_3^- is 8.121×10^{-7} and 8.207×10^{-7} in surface and bottom water samples, respectively. Moreover, ~ 50% of water samples from both levels show that the CR value for adults exceeded its permissible limits (1×10^{-6}). Besides, ~57.14% of the surface and 50% of the bottom water samples show that the CR values for children exceeded their limits. And for babies, 21.43% of surface water samples and 28.57% of bottom water exceeded their permissible limits. Moreover, Seal et al. (2022) pointed out the higher carcinogenic risk in the Damodar River Basin and recognized that the cities of Durgapur and Burnpur-Asansol are major sources of CR. The above results also explain that the drinking water quality is poor, leading to major health risks and poisoning if consuming this water without any purification and treatment. NO_3^- toxicity leads to methemoglobinemia in pregnant women and infants, and it also leads to stomach cancer, respiratory problems, headaches and drowsiness, thyroid gland hyperplasia, and multiple sclerosis (Duvva et al., 2022; Jannat et al., 2022).

Therefore, with the increased agricultural and industrial activities, nitrogen pollution in water resources is also increasing exponentially (Zhang et al., 2018). The nitrogen fertilizer used in agricultural land produces water-soluble nitrate and nitrite compounds that move through surface runoff, groundwater, and rivers (Shukla & Saxena, 2020). In the context of DRB, surface runoff from agricultural fields (on-bed agriculture, especially on the Damodar chars or river islands and off-bed agriculture) poses threats to the future due to the extensive use of nitrogen fertilizer. Nitrogen is one of the chief components of the ecosystem and is also essential for the human body; however, exceeding the maximum contaminant level poses serious public health risks. Furthermore, when nitrate (NO_3^-) converts into nitrite (NO_2^-), it leads to several diseases such as cancer, hypertension, and Blue-baby syndrome (Li et al., 2019). High consumption of nitrate could be the reason for methemoglobinemia in infants (below 3 months), due to which the blood does not carry and release oxygen, leading to breath shortness, susceptibility to illness, cardiovascular problems, and even death due to asphyxiation in high concentrations (Bazeli et al., 2020). Water that is contaminated with nitrate, if consumed by a pregnant woman during the pregnancy period, results in adverse pregnancy outcomes (Ward et al., 2005). Thus, nitrate and nitrite compounds are highly responsible for different types of carcinogenic risks, including brain cancer, stomach cancer, bladder cancer, kidney cancer, colorectal cancer and pancreatic cancer (Picetti et al., 2022; Ward et al., 2005).

4.4 Alteration of Ecological Water Quality

All organisms depend on water as it is a crucial natural resource (Karataş & Karataş, 2016). Throughout the world, and water-related problems persist, which is a major issue for all (Islam et al., 2025; Aboyeji, 2013). In the present day, people are aware of it and they are concerned about surface and groundwater pollution as water is required for various purposes and environmental contexts (Inyinbor Adejumoke et al., 2018; Jayaswal et al., 2018). The freshwater ecosystem is controlled by the quantity and quality of water as well as seasonality (Baron et al., 2002; Hoque et al., 2022a). The quality and quantity of river water are controlled by several natural and anthropogenic factors, such as surface and sub-surface runoff, rock weathering, precipitation, temperature, mining activities (Bhateria & Jain, 2016), urban-industrial and agricultural effluents (Al-Hadithi et al., 2019), and structural interventions (e.g., dam, barrage weir, rail, and road bridge) (Islam et al., 2023; Hoque et al., 2022b). Besides, the diversion and interruption of river water (Hoque et al., 2024), excess water extraction for irrigation, industrial and domestic purposes highly affect the pollution level in river water (Falowo et al., 2017), resulting in environmental degradation and loss of riverine ecosystem as well as biota (Hoque et al., 2022a).

Due to higher pollution levels, reflected through higher BOD and lower DO concentration, the mixing of urban-industrial effluents and agricultural runoff with river water results in high concentrations of algae, moss, freshwater sponges, crustaceans, insects, and gastropods, degrading the habitat productivity of the river (Leal et al., 2021). Hence, the assessment of the spatio-temporal quality of river water is essential for numerous purposes. To evaluate the water quality, the Vietnam Environmental Agency (VEA) developed a water quality index (WQI) (Son et al., 2020; Ha et al., 2021). Furthermore, to measure the river water pollution level, comprehensive and organic pollution, eutrophication, and bacteriological water quality are required to be evaluated.

Furthermore, this study presents few major objectives that are—(1) to measure the ecological, bacteriological, and overall water quality of the Damodar River, (2) to identify variations in river water in terms of spatial and temporal quality with the help of monthly water quality data obtained from the West Bengal Pollution Control Board (WBPCB, 2022) for 11 monitoring stations on the Damodar River. The data for water quality were collected from February 2014 to January 2023 (Fig. 5.1; Table 5.1) for 11 stations. The organic pollution index (OPI) and eutrophication index (EI) are used to assess the ecological water quality of the Damodar River. Total coliforms (TC), faecal coliform (FC), and faecal streptococci (FS) concentrations are utilized to analyse the bacteriological water quality of the river. Furthermore, the overall water quality has been determined using the water quality index (WQI) and comprehensive pollution index (CPI).

4.4.1 Ecological Water Quality

4.4.1.1 Organic Pollution Index

In order to measure organic pollution levels for different stretches of the Damodar River, OPI has been employed. The present study affirms that the average OPI score ranges from 3.61 to 15.77, having a mean value of 4.96 for pre-monsoon, 3.64 to 6.92 with a mean value of 4.46 for monsoon, and 3.65 to 7.83 with a mean value of 4.34 in the post-monsoon period (Table 4.S7). It is also determined that the highest

OPI score is at station 10 in all the respective seasons among the other stations, but within the station, it is high in the pre-monsoon. The lowest OPI score is detected at station 4 in all the seasons. The spatio-temporal map (Fig. 4.3a) displays the variation in OPI scores, varying from 3 to 4 in the upper part of the concerned river in the pre-monsoon at stations 2–8, in monsoon at stations 2, 4, and 8, and in post-monsoon at stations 2, 4, 7 and 8 pointing out that above-mentioned sections have moderate organic pollution in respective seasons based on the categorical classification of OPI. The middle and lower part of the concerned river contains the OPI score within >4 in all the seasons, while the upper part in the pre-monsoon at stations 1, in monsoon at stations 1, 3, 5, 6, and 7, and post-monsoon at stations 1, 3, 5, and 6, indicating heavy organic pollution.

Furthermore, depending on the categorical classification of the OPI score, it is determined that only 2.78% of water samples began to be contaminated ($1 < OPI \leq 2$ score) in the monsoon at station 6 and post-monsoon at station 8. At station 10, higher percentage of water samples fell into >4 of OPI score, i.e., high organic

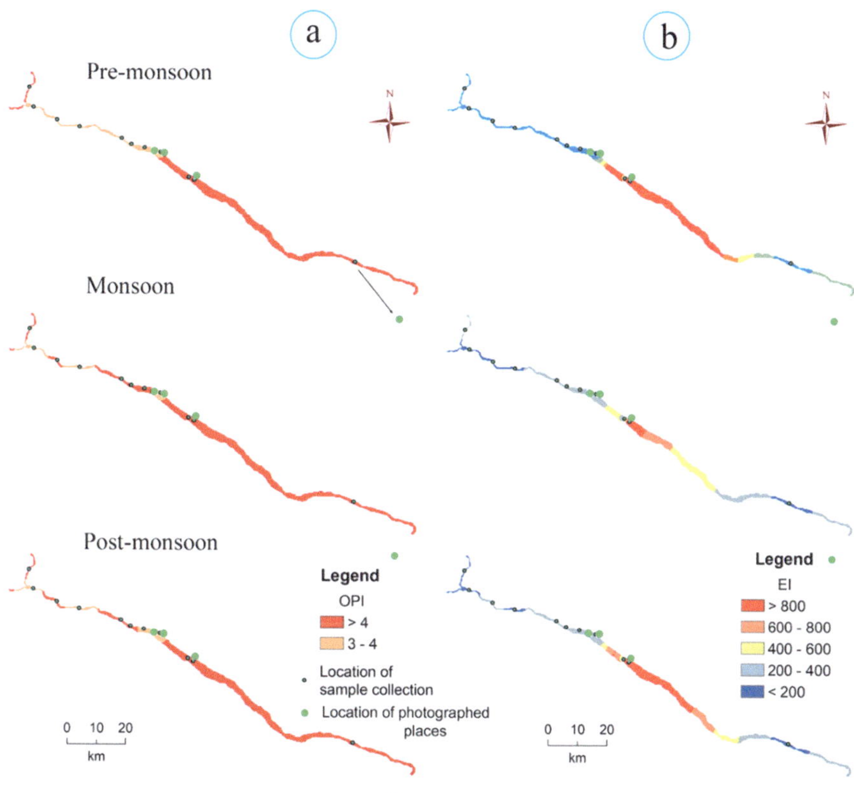

Fig. 4.3 Spatio-temporal variation of ecological water quality; (**a**) OPI and (**b**) EI in water. (Note: The south-east directed black arrow indicates the location of photographed places 5 and 6, which are located 30 km downstream from the monitoring station 11, i.e., water intake point at Bardhaman)

pollution category in pre-monsoon (94.29%), monsoon (100.00%), and post-monsoon (100.00%) seasons, however, a low percentage of water samples fell into this group at station 7 (25.71%) in pre-monsoon, station 4 (19.44%) in monsoon and station 4 (33.33%). In the category of light organic pollution ($2 < OPI \leq 3$) class, the highest percentage of water samples is found to be at station 4 (25.71%) in pre-monsoon, 8 (36.11%) in monsoon, and 4 (25.00%) in post-monsoon seasons. Moreover, this category contains the lowest percentage of water samples that are detected at station 2 in pre-monsoon (5.71%), stations 2 and 3 in monsoon (5.56%), and station 6 in post-monsoon (8.33%) seasons.

Moreover, in comparison to all categories, a higher percentage of water samples is observed in the classification group of moderately organic pollution ($3 < OPI \leq 4$) among the other classification groups in all the respective seasons. 71.73% percentage of water samples fell into this group ($3 < OPI \leq 4$) at station 3 (in pre-monsoon, 1 and 3 (61.11%) in monsoon, and 2 (61.11%) in post-monsoon seasons whereas, the lowest percentage of water samples is found to be in this group at station 10 in pre-monsoon (5.71%), at station 8 in monsoon (33.33%) and station 5 and 8 in post-monsoon (36.11%) (Table 4.S8). The above-mentioned results demonstrate that in the pre-monsoon, organic pollution is high, whereas in the post-monsoon it is low, and again it is high after the Tamla *Nallah* confluence in all the seasons.

4.4.1.2 Eutrophication Index

The score for EI has also been calculated and evaluated to examine the eutrophication level in the Damodar River water from the perspective of spatio-temporal dynamics. It is found that the EI score is high in all the seasons and also at all the stations in the spatio-temporal dimension, denoting high eutrophication in river water. The average EI score of 572.84, 343.14, and 447.12 is measured in pre-monsoon, monsoon, and post-monsoon, respectively. At station 4, the average lowest EI score is (85.61) in pre-monsoon, 2 (101.37) in monsoon, and also 4 (153.08) in the post-monsoon period; however, the highest score is determined in pre-monsoon (5106.27), monsoon (1298.20), and post-monsoon (2655.13) at station 10. The coefficient of variation (CV) is very high for all the stations in all the seasons as per the calculations and findings (Table 4.S9).

Besides, the map that represents spatio-temporal variation indicates that the EI score >800 is found to be in the middle part of the Damodar River water; whereas, a score <200 is found to be in the upper part in pre-monsoon and uppermost part in pre-and post-monsoon and some portion of the lower part near at station 11 in all the seasons (Fig. 4.3b). In the monsoon and post-monsoon seasons, the EI score ranges from 200 to 400 in the upper and lower parts of the river (Fig. 4.3b). Moreover, 93.25%, 90.15%, and 94.04% of water samples in pre-monsoon, monsoon, and post-monsoon, respectively, fell into EI >9 categories, determining extreme eutrophication in all seasons in the Damodar River water (Table 4.S10). Only a few percentages, i.e., 5.71%, 5.71% 2.86%, 2.86%, 2.86%, 5.71%, and 2.86% of water samples at stations 2, 4, 6, 7, 8, 9, and 11 in pre-monsoon, 2.78%, 2.78%, 2.78%, 5.56%, 2.78%, 2.78%, and 8.33% of water samples at stations 1, 2, 3, 4, 7, 9, and 11 in monsoon, and 2.86% and 5.71% of water samples at stations 2 and 7 in post-monsoon, respectively, range from $0 \leq EI \leq 1$ depicts that less chances of eutrophication based on categorical classification (Chen, 2021). The above-mentioned findings explain that the level of eutrophication is high in the pre monsoon, having a mean of 572.84; however, the monsoon period experiences low eutrophication with a mean of 343.14, which is also high after the Tamla *Nallah* confluence, i.e., station 10 (EI = 5106.27 in pre-monsoon, EI = 1298.20 in monsoon and EI = 2655.13 in post-monsoon) in all the seasons. In addition, a very high EI score is found at all the stations in all seasons, depicting very high or extreme eutrophication.

4.4.2 *Bacteriological Water Quality*

4.4.2.1 Total Coliform

In the present work, it is exposed that in Damodar River water, the average concentration of total coliform organisms or bacteria ranges from 6942.86 to 17,226.47, 11,313.89 to 33,208.33, and 4588.89 to 14,813.89 MPN/100 mL, having a mean value of 11,315.13, 19,326.52, and 10,270.48 MPN/100 mL in the pre-monsoon, monsoon, and post-monsoon period, respectively. In the pre-monsoon period, the highest average TC concentration is detected (17,226.47 MPN/100 mL) in this river water at station 10 and the monsoon period (33,208.33 MPN/100 mL) and post-monsoon period (14,813.89 MPN/100 mL) at station 1 whereas in pre-monsoon the lowest TC concentration was detected (6942.86 MPN/100 mL) at station 4, monsoon (11,313.89 MPN/100 mL) at station 11 and in post-monsoon (4588.89 MPN/100 mL) at station 3 (Table 4.S11). Additionally, the spatial variation map and mean values of TC concentration reveal that the water quality is good in the post-monsoon period in comparison to other seasons. In the upper section concentration of TC ranges from 20,000 to 40,000 MPN/100 mL during the monsoon at stations 1, 2, and 7. While in the upper section, TC concentration lies <10,000 MPN/100 mL in the pre- and post-monsoon, and the lower section, also the pre-monsoon season at station 11, follows the same category (Fig. 4.4a). Besides, based on the environmental standard (5000 MPN/100 mL) the TC has been classified and it was revealed that 68.57%, 60.00%, and 65.00% of samples of stations 1, 6, and 10 in pre-monsoon, 61.11% and 80.56% of water samples of station 1 and 10 in monsoon and 55.56% of water samples of station 10 in post-monsoon exceeded its environmental standard (Tables 4.S11, S12 and S14). The above explanation affirms that in the monsoon, TC concentration is high, and in the post-monsoon period, it is low. Nevertheless, it recognizes that at station 1, TC concentration is high in monsoon and post-monsoon, and also it is higher near Mujher Mana village, which is situated downstream of the Durgapur barrage in the pre-monsoon season.

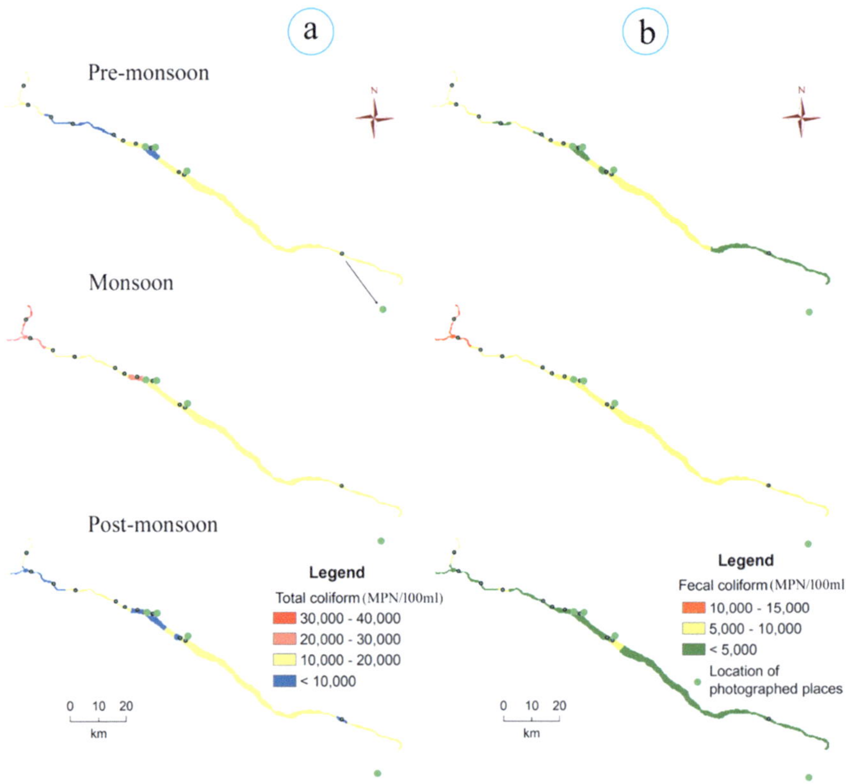

Fig. 4.4 Spatio-temporal variation of bacteriological water quality; (**a**) TC and (**b**) FC concentration in water samples. (Note: The south-east directed black arrow indicates the location of photographed places 5 and 6)

4.4.2.2 Foecal Coliform

The analysis for FC concentration reveals, it was detected that in the monsoon season average FC concentration is high, following pre- and post-monsoon, which range from 3494.29 to 9354.29 (mean 5675.58) MPN/100 mL, 5005.56 to 13,863.89 (mean 8510.10) MPN/100 mL and 2900.00 to 5827.78 (mean 4275.71) MPN/100 mL, having a higher CV in pre-monsoon, monsoon and post-monsoon seasons accordingly. At station 9, the lowest FC concentration is detected in the pre-monsoon, 11 in monsoon, and 3 in post-monsoon seasons, whereas at station 10, the highest FC concentration is noticed in the pre- and post-monsoon seasons and at station 1 in the monsoon season (Table 4.S15).

The spatial variation map displays that the distribution of FC <5000 MPN/100 mL covers the upper and lower stretches in pre-monsoon and almost all stretches, excluding station 10 in post-monsoon. The concentration of FC from 5000–10,000 MPN/100 mL is found to be in the middle and upper stretches in pre-monsoon, almost all stretches, except stations 1 and 2 in monsoon, and in the post-monsoon season at station 10. The concentration of FC ranges from 10,000 to 15,000 MPN/100 mL in the monsoon at the topmost zone of the upper stretch (Fig. 4.4b). Moreover, based on the environmental standard (1000 MPN/100 mL), the water quality of the Damodar River has been classified and it is observed that 91.69%, 96.97%, and 93.18% of water samples in the pre-monsoon, monsoon, and post-monsoon seasons, respectively, surpassed the permissible limit (Tables 4.S11, S12 and S14). Furthermore, for the bathing standard limit (2500 MPN/100 mL), it is observed that 65.97%, 69.95%, and 51.52% of water samples in pre-monsoon, monsoon, and post-monsoon, respectively, surpassed the permissible limit. The outcome affirms that the water of the Damodar River is not suitable for bathing purposes. In addition, the monsoon period has a higher FC concentration, whereas in the post-monsoon period, it is low. It is also detected that, near the downstream of Durgapur barrage, close to Mujher Mana village, the FC concentration is high in the pre- and post-monsoon, whereas in the monsoon season, it is high at station 1.

4.4.2.3 Faecal Streptococci

Streptococci originated from humans, which is evident in pollution. Some bacteria from animal faeces or plants are also categorized in this assembly of Streptococci (Mezbour et al., 2018). Depending on the data that is available monthly, it is determined that the average concentration of FS in Damodar River water ranges from 16.14 to 29.57, 16.75 to 25.50, and 15.91 to 23.18 MPN/100 mL, having a mean value of 20.92, 19.92, and 18.91 MPN/100 mL in the pre-monsoon, monsoon, and post-monsoon periods, respectively (Table 4.S16). At station 9, the highest average concentration of FS is observed in pre-monsoon, and at station 10 in monsoon and post-monsoon seasons; however, the lowest value is observed at stations 3, 4, and 11 in pre-monsoon, monsoon, and post-monsoon periods, respectively, in the concerned study area. The above-mentioned outcomes pointed out that the concentration of FS is high in the pre-monsoon whereas low in the post-monsoon period. It is also found that the concentration of FS is noticed to be higher at the downstream of the Durgapur barrage close to the village Mujher Mana regardless of seasons (Fig. 4.5a).

4.4.3 Bathing Water Quality

Moreover, based on the environmental or bathing standard limit (500 MPN/100 mL), the water quality of the Damodar River has been classified, and it was found that water samples for all stations are below the standard permissible limit, making the water suitable for bathing purposes. Additionally, the water quality for bathing purposes of the river has also been evaluated and observed that it ranges from 0.91 to 1.58, having a mean value of 1.15 in the pre-monsoon, 1.07–1.95 with a mean value of 1.43 in monsoon and 0.82–1.22 having a mean value of 0.98 in the post-monsoon

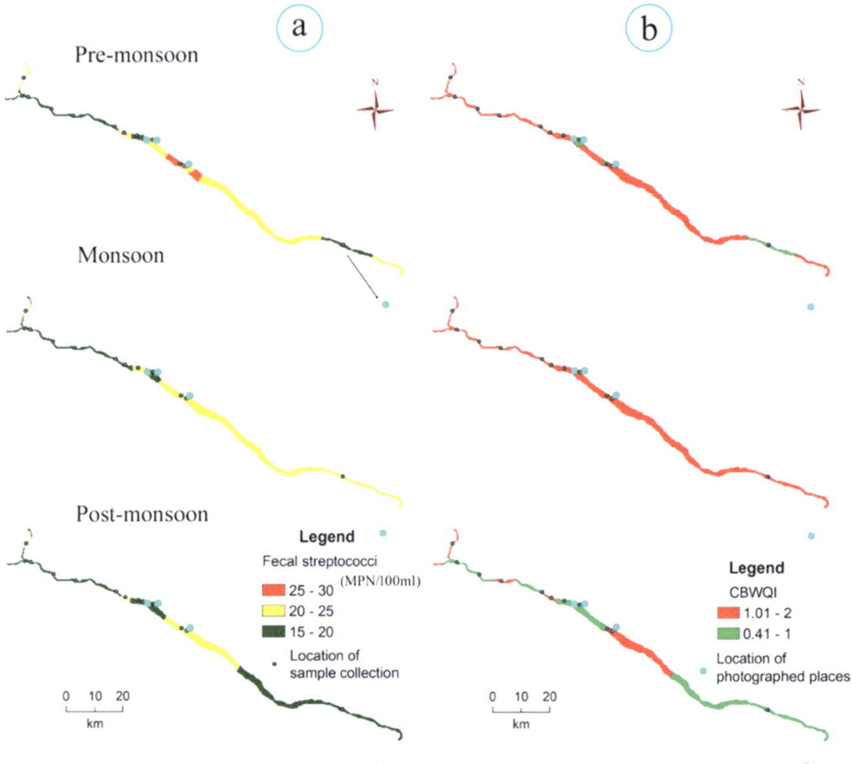

Fig. 4.5 Spatio-temporal variation of (**a**) bacteriological water quality as FS concentration and (**b**) comprehensive bathing water quality index (CBWQI). (Note: The south-east directed black arrow indicates the location of photographed places 5 and 6)

season (Table 4.S17). Furthermore, the map for spatio-temporal variation displays that the CBWQI score >0.41–2 covers the entire stretch of the river. However, the pollution level in the middle stretch is higher than the upper and lower stretches and the water during the monsoon season is unsuitable for bathing uses (Fig. 4.5b).

4.4.4 Overall Water Quality

4.4.4.1 Overall Water Quality Index

The overall quality of water has been evaluated with the help of VEA-recommended WQI for several needs for this study, and it was observed that the values of average WQI ranged from 49.80 to 79.83, 39.67 to 63.22, and 50.17 to 74.27 with a 69.25, 54.87, and 67.77 mean value in pre-monsoon, monsoon and post-monsoon seasons, accordingly (Table 4.S18). Good water quality is depicted by a higher WQI score and vice versa. At station 10 (i.e., Damodar after Tamla *Nallah* confluence), the lower WQI values having higher CV were measured for all the seasons. At station 2, the higher WQI values were detected during pre-monsoon, 11 in monsoon, and 7 in post-monsoon. In the monsoon season, the lowest CV was found following post-monsoon (Table 4.S18), indicating that the variation in data for the WQI score was highest in the monsoon and lowest in the pre-monsoon season.

Furthermore, the map for spatial variation (Fig. 4.6a) shows that at stations 2, 3, 7, and 8, the WQI score is high, which are situated at the upper stream of Durgapur barrage, indicating good quality water for domestic purposes in the pre-monsoon. While at station 10, lower WQI values are found for all seasons, whereas similar outcomes are noticed for station 1, indicating poor water quality only in the monsoon period, i.e., water can be utilized for transportation and equivalent purposes only. The rest of the stations fall under the moderate category of water quality regardless of seasons, indicating that in these sections water can be utilized for irrigation and equivalent purposes according to the categorical classification of the WQI score (Table 4.S19).

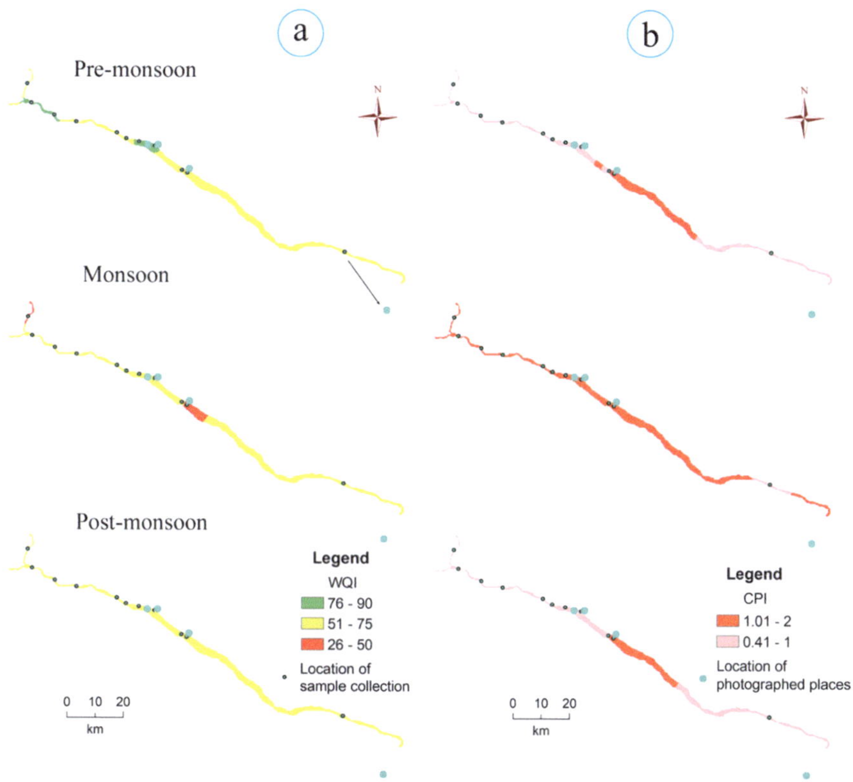

Fig. 4.6 Spatio-temporal variation of overall water quality; (**a**) WQI and (**b**) CPI. (Note: The south-east directed black arrow indicates the location of photographed places 5 and 6)

In addition, the score for WQI has been classified depending on the categorical classification (Table 4.S19) and determined that the 37.14% of water samples which is highest of station 1 in pre-monsoon, 41.67% of water samples of stations 1, 2, 3, and 10 in monsoon, and 33.33% of water samples of station 3 in post-monsoon fall under the category of heavy water pollution (WQI score 0–25). Whereas, at stations 2 and 8 in pre-monsoon, the lowest percentage of water samples (11.43%), at stations 8 and 9 in monsoon (25%), and at stations 2 and 7 in post-monsoon (16.67%) fell under the category of heavy pollution. A very small amount or percentage of water samples in all the respective seasons and stations, excluding station 10, fall into the category of poor water quality. The highest percentage 42.86% of the water sample at station 10 in pre-monsoon, 44.44% at station 8 in monsoon, and 30.56% at station 10 in post-monsoon fell into the moderate category however, the lowest, 5.71% at stations 4 and 7 in pre-monsoon, 13.89% at stations 2 and 3 in monsoon and 5.56% in post-monsoon at station 3 lies below this category. In the good category, the highest percentage, 48.57% in pre-monsoon was found at station 2, 33.33% in monsoon at stations 1–4, and 52.78% in post-monsoon at station 5, whereas the lowest percentage, 14.29%, 11.11%, and 22.22% of water samples at station 10 in all the seasons fell into this category. The last category that is very good water quality, pre-monsoon season is having lowest percentage of water samples (2.86%), monsoon (0.00%), and post-monsoon (0.00%) at station 10 fell into this category however, the highest percentage, 42.86% in pre-monsoon at station 3 and 7, at stations 4, 8 and 9 13.89% in monsoon and 33.33% of water samples at stations 4 and 9 in post-monsoon fell into this category of very good water quality (Table 4.1).

According to the average WQI score, these results show that during monsoon, the water quality is most unacceptable, followed by post-monsoon and pre-monsoon in the temporal dimension, while, throughout the year, as per comparison, most unacceptable water quality found at the Damodar River immediately downstream of Durgapur barrage or after the Tamla *Nallah* confluence among the other stations in the spatial variation. The maximum water sample, 28.57% in pre-monsoon, 41.67% during monsoon, and 27.78% during post-monsoon, after the Tamla *Nallah* confluence and water samples in post-monsoon at Barakar, Dishergarh, and IISCO were also found to have highly polluted water, i.e., 41.67%. However, during the monsoon season, the Tamla *Nallah* confluence and Barakar are found to have poor water quality. Moreover, during pre-monsoon at Dishergarh, IISCO and Andal, good quality of water was found, whereas, based on WQI average score, the moderate quality of water was found throughout the year in the remaining parts.

4.4.4.2 Comprehensive Pollution Index

As per the comprehensive study of pollution through the spatio-temporal dynamic, the contamination level in the Damodar River has been measured to evaluate the average pollution level during monsoon is 1.23, and was maximum, followed by 0.94 during pre-monsoon and 0.88 during post-monsoon. The mean CPI score ranges between 0.75 and 1.75 during pre-monsoon, 0.98 and 1.67 during monsoon,

Table 4.1 Season-wise classification of water samples (percentage) in different monitoring stations based on WQI score

Seasons	WQI classification	Stations										
		1	2	3	4	5	6	7	8	9	10	11
Pre-monsoon	0–25	37.14	11.43	17.14	22.86	22.86	22.86	17.14	11.43	22.86	28.57	22.86
	26–50	–	–	–	–	–	–	–	–	2.86	11.43	2.86
	51–75	11.43	8.57	14.29	5.71	14.29	22.86	5.71	14.29	14.29	42.86	22.86
	76–90	40	48.57	25.71	40	40	37.14	34.29	45.71	37.14	14.29	34.29
	91–100	11.43	31.43	42.86	31.43	22.86	17.14	42.86	28.57	22.86	2.86	17.14
Monsoon	0–25	41.67	41.67	41.67	27.78	38.89	33.33	33.33	25	25.00	41.67	19.44
	26–50	–	–	2.78	2.78	2.78	2.78	5.56	2.78	2.78	22.22	8.33
	51–75	25	13.89	13.89	22.22	25	25	27.78	44.44	27.78	25	38.89
	76–90	33.33	33.33	33.33	33.33	30.56	30.56	22.22	13.89	30.56	11.11	25
	91–100	–	11.11	8.33	13.89	2.78	8.33	11.11	13.89	13.89	–	8.33
Post-monsoon	0–25	30.56	16.67	33.33	22.22	22.22	25	16.67	19.44	19.44	27.78	13.89
	26–50	2.78	2.78	2.78	–	–	2.78	–	–	–	19.44	–
	51–75	8.33	16.67	5.56	13.89	11.11	16.67	13.89	19.44	19.44	30.56	25
	76–90	41.67	36.11	36.11	30.56	52.78	38.89	44.44	36.11	27.78	22.22	38.89
	91–100	16.67	27.78	22.22	33.33	13.89	16.67	25	25	33.33	–	22.22

Computed by the authors, 2023

and 0.71 and 1.69 during the post-monsoon season. The highest CPI score is observed at station 10 throughout the year, whereas the lowest CPI score is found at station 4 during pre-monsoon, 11 during monsoons, and 3 during post-monsoon season. In the spatial dimension, the highest CV score is observed as 113.60% and 98.43% at station 2 during pre-monsoon and 99.04% during post-monsoon at station 10, whereas the lowest CV score is observed at station 5 during pre-monsoon, 11 during monsoon and 7 during post-monsoon. Furthermore, a mean higher CV is found during the monsoon, followed by the pre- and post-monsoon seasons in temporal dynamics (Table 4.S20).

The quality of water during the monsoon season is also more unacceptable than the remaining seasons, as shown in the spatial variation map (Fig. 4.6b). The quality of water varies between 1.01 and 2 during the monsoon season in almost all the regions or rivers, except station 11, which lies in a medium pollution level according to the CPI score's categorical classification. Furthermore, the river water quality of the upper and lower parts, except the middle section or the downstream region of Durgapur barrage (station 10) of the study area, ranging between 0.41 and 1 during the pre- and post-monsoon season, shows less pollution (Fig. 4.6b).

Moreover, according to the CPI score, the water quality has been categorized and observed that 2.86% of water samples ranged between 0.21 and 0.40, which is sub-clean at stations 2 during pre-monsoon, 4 during monsoon, and 3 and 7 during the post-monsoon period only. Therefore, 80.52% of water samples during pre-monsoon, 59.09% of water samples during monsoon, and 80.56% of water samples during post-monsoon periods are found in 0.41–1.00 which is moderately polluted, high at station 9 (94.29%) and 8 (94.29%) during pre-monsoon, 9 (72.22%), 8 (72.22%), and 4 (72.22%) during monsoon, 11 (91.67%), 2 (91.67%), and 9 (91.67%) during the post-monsoon period. The highest % of water samples at station 10 ranges between 1.01 and 2.00, which is moderately polluted during pre-monsoon (51.43%), monsoon (59.44%), and post-monsoon (55.56%) periods only. Furthermore, during pre-monsoon and post-monsoon, the water sample exhibits 17.14% and 13.89% at station 10 and at station 1, 16.67% of water samples are found as >2.01 CPI score category, which is strongly polluted and is also high over the rest of the period (Table 4.2). During the monsoon period, the CPI results show that the pollution level is also high in temporal dynamics and also high at Durgapur barrage' lower stretch after the Tamla *Nallah* confluence near the village of Mujher Mana over the remaining stations throughout the year in terms of comprehensive pollution in the spatial dimension.

4.4.5 Spatio-temporal Variations in Water Quality

ANOVA results show that there is no significant difference in TC and FC across all seasons, FS in the monsoon and post-monsoon seasons, and OPI and CPI in the monsoon within the spatial dynamics. However, other indices exhibit significant spatial variation during the remaining seasons (Table 4.3). This indicates a notable spatial variation in water quality. Furthermore, ANOVA analysis of the temporal

Table 4.2 Season-wise classification of water samples (percentage) in different monitoring stations based on CPI score

Seasons	CPI classification	Stations										
		1	2	3	4	5	6	7	8	9	10	11
Pre-monsoon	0.21–0.40	–	2.86	–	–	–	–	–	–	–	–	–
	0.41–1.00	71.43	85.71	85.71	85.71	91.43	82.86	85.71	94.29	94.29	31.43	77.14
	1.01–2.00	28.57	2.86	11.43	14.29	8.57	11.43	8.57	–	2.86	51.43	20
	>2.01	–	8.57	2.86	–	–	5.71	5.71	5.71	2.86	17.14	2.86
Monsoon	0.21–0.40	–	–	–	2.78	–	–	–	–	–	–	–
	0.41–1.00	47.22	63.89	69.44	72.22	61.11	52.78	58.33	72.22	72.22	16.67	63.89
	1.01–2.00	36.11	25	19.44	19.44	27.78	33.33	30.56	19.44	19.44	69.44	30.56
	>2.01	16.67	11.11	11.11	5.56	11.11	13.89	11.11	8.33	8.33	13.89	5.56
Post-monsoon	0.21–0.40	–	–	2.78	–	–	–	2.78	–	–	–	2.78
	0.41–1.00	80.56	91.67	86.11	77.78	83.33	77.78	88.89	86.11	91.67	30.56	91.67
	1.01–2.00	13.89	5.56	11.11	19.44	13.89	19.44	8.33	11.11	8.33	55.56	–
	>2.01	5.56	2.78	–	2.78	2.78	2.78	–	2.78	–	13.89	5.56

Computed by the authors, 2023 (Note: Any water sample lies in the 0–0.20 CPI class category)

Table 4.3 One-way ANOVA exhibiting the spatio-temporal variation in water quality

		Spatial*				Temporal**			
Indicators/indices	Seasons	*F*	*P*-value	*F*-critical	Remarks	*F*	*P*-value	*F*-critical	Remarks
OPI	Pre-monsoon	7.32	1.29E – 10	1.86	Alternative	1.73	0.177753	3	Null
	Monsoon	1.80	0.058429	1.86	Null				
	Post-monsoon	5.94	2.26E – 08	1.86	Alternative				
EI	Pre-monsoon	5.00	7.78E – 07	1.86	Alternative	0.73	0.486455	3	Null
	Monsoon	5.67	6.09E – 08	1.86	Alternative				
	Post-monsoon	6.22	7.86E – 09	1.86	Alternative				
TC	Pre-monsoon	0.62	0.799913	1.86	Null	12.12	6.18E – 06	3	Alternative
	Monsoon	0.94	0.492111	1.86	Null				
	Post-monsoon	0.76	0.663003	1.86	Null				
FC	Pre-monsoon	1.11	0.355525	1.86	Null	16.51	8.53E – 08	3	Alternative
	Monsoon	1.12	0.348389	1.86	Null				
	Post-monsoon	1.15	0.326891	1.86	Null				
FS	Pre-monsoon	2.65	0.008733	1.98	Alternative	1.99	0.138914	3.02	Null
	Monsoon	1.18	0.320146	1.96	Null				
	Post-monsoon	1.39	0.194878	1.92	Null				
WQI	Pre-monsoon	3.42	0.000266	1.86	Alternative	28.75	6.44E – 13	3	Alternative
	Monsoon	1.91	0.043018	1.86	Alternative				
	Post-monsoon	2.06	0.026552	1.86	Alternative				
CPI	Pre-monsoon	4.86	1.36E – 06	1.86	Alternative	20.43	1.89E – 09	3	Alternative
	Monsoon	1.78	0.062502	1.86	Null				
	Post-monsoon	6.72	1.19E – 09	1.86	Alternative				

Computed by the authors, 2023 (Note: Degree of freedom: 10 for spatial and 2 temporal variations, significance level: 0.05; Note: * indicates the spatial variation of water quality among the different monitoring stations in a particular season; ** indicates the temporal variation of water quality from the initial years (2014) to the penultimate year (2023)) variation of water quality indices reveals significant differences in overall water quality indices (WQI and CPI), whereas there is no significant difference in ecological water quality indices (OPI and EI). Additionally, bacteriological water quality indicators like TC and FC show significant differences over time, while FS does not (Table 4.3). This suggests a considerable temporal variation in the water quality of the Damodar River. Overall, these ANOVA results demonstrate that there are significant spatio-temporal variations in the river's water quality.

4.5 Alteration of Groundwater Quality of the Damodar River Basin

The groundwater quality in the Damodar River Basin has been significantly altered due to various anthropogenic activities (industrial, agricultural and mining) and natural processes, leading to pollution and resource depletion. This alteration leads to a deterioration of water quality for drinking and other purposes. Pollutants such as arsenic, fluoride, and heavy metals are introduced through agricultural practices, industrial effluents, and urbanization, posing health risks. Besides, overexploitation of groundwater for irrigation and other uses also contributes to the problem, affecting both the quantity and quality of available water in the basin area.

The basin is characterized by large-scale industries, densely populated urban areas, extensive mining activities and intensive agricultural practices, which contribute to both surface and groundwater pollution. Urban-industrial effluents with disposal, agricultural effluent and mining operations introduce heavy metals, chemicals, and other pollutants into the water system. Besides, natural geological processes also lead to groundwater contamination, with arsenic and fluoride being common geogenic elements found in the aquifers of the basin area.

The alteration of groundwater quality impacts public health and the sustainability of water resources in the basin area. Several studies have shown that the basin's groundwater contains high levels of contaminants such as arsenic, fluoride, and heavy metals, exceeding permissible limits. Arsenic and fluoride elements can contaminate groundwater through natural mineralization processes and agricultural activities, posing health risks to local populations. Industrial and mining activities discharge heavy metals into the river and its tributaries, pollute groundwater and pose a health risk to the population.

4.6 Natural and Anthropogenic Drivers of Qualitative Environmental Flow Alterations

4.6.1 Climate Change

Like quantitative environmental flow alteration, climate change plays a crucial role in altering qualitative environmental flows through changes in precipitation patterns, increased evaporation, and variations in temperature and water availability.

Climate change results in more frequent and intense rainfall events (floods) or prolonged periods of drought, which affect the quality, quantity, and timing of flow in aquatic systems. Higher precipitation leads to increased sediments or turbidity in water bodies. A high concentration of turbidity in water renders it unsuitable for drinking. For instance, the water quality of the Damodar River during the monsoon period is particularly unacceptable due to elevated turbidity levels. Droughts concentrate pollutants (salts, heavy metals, nutrients) in water, deteriorating its quality. However, extended periods of drought lead to diminished concentrations of essential nutrients in aquatic ecosystems due to little or no flow. Rising temperatures from climate change increase evaporation rates and reduce water availability in aquatic systems, causing thermal stress in these habitats, as warmer water holds less oxygen, which poses risks to aquatic life (e.g., fish and macroinvertebrates), especially under low-flow conditions. Increasing temperatures alter species composition, favouring tolerant species over susceptible natives. The rise in temperature and reduction in water availability impact ecosystem functions and processes such as photosynthesis, water use, and plant growth. Changes in water flow and temperature modify the physical, chemical, and biological properties of water, influencing water quality and the ability of aquatic ecosystems to thrive. These alterations can disrupt natural ecosystems, change habitat suitability, and affect various aspects of water resource management.

4.6.2 LULC Changes

Land Use and Land Cover (LULC) changes significantly alter the qualitative environmental flow by impacting the water cycle and influencing factors such as infiltration, evaporation, and runoff. These changes lead to alterations in water availability, nutrient loads, and sediment transport, affecting water quality in aquatic systems. Consequently, these alterations result in water pollution, impacting its chemical and biological quality. Intensive agricultural practices introduce agricultural effluents into river systems, carrying excess nutrients (like nitrogen and phosphorus) and pesticides, which can lead to algal blooms and oxygen depletion in aquatic ecosystems. Urban-industrial effluents contain heavy metals, pathogens, and other pollutants from various sources that are discharged into river systems. Extensive mining activities in the basin area introduce numerous pollutants into the river systems, which alter the physico-chemical properties of water. Deforestation increases soil erosion, leading to sedimentation and increased turbidity in water, while reducing water infiltration, negatively affecting water quality and harming aquatic life. Hence, changes in LULC disrupt ecosystem services, including water purification, reduce biodiversity, and affect the overall health of aquatic ecosystems.

4.6.3 Emerging Contaminants' Role

Emerging pollutants such as pesticides and pharmaceuticals alter the qualitative environmental flows by changing the physico-chemical properties of water, which affects aquatic ecosystems. These pollutants decrease dissolved oxygen, promote toxic algal blooms, and influence aquatic organisms' behaviour and reproductive success, leading to migration and extinction. Emerging pollutants are introduced into river systems through various sources, including industrial discharge, agricultural runoff, wastewater, mining activities, improper waste disposal, and household and personal products. Industrial processes release a large amount of chemicals into the environment, some of which exist for long periods and significantly impact ecosystems. Besides, extensive use of pesticides and fertilizers in agricultural fields introduces contaminants into water through runoff, leaching, and other mechanisms. Moreover, discharging wastewater from sewage treatment plants and industrial facilities releases pharmaceuticals, personal care products, and other chemicals into river systems. The Damodar River Basin is the storehouse of numerous minerals. Hence, mining activities introduce various pollutants into the river systems. Additionally, the use of personal care products, household cleaners, and pharmaceuticals contributes to the presence of emerging contaminants in wastewater and other aquatic systems. Illegal dumping and improper waste disposal can introduce pollutants into the environment, especially in urban areas.

4.6.4 Nutrient Loss Due to Dam Construction

Dam construction potentially affects nutrient flow in river systems, reducing nutrient transport downstream and disrupting ecosystems. It acts as a barrier to the natural flow of nutrients, trapping them in the reservoir and altering the balance of nutrients reaching coastal areas. It reduces nutrient supply downstream and increases nutrient concentrations within the reservoir, contributing to eutrophication and harming the reservoir's ecosystem. Consequently, it alters the nutrient ratios in water systems and favours some algae species over others, resulting in ecosystem imbalance and potentially harmful algal blooms.

4.7 Improvement of Qualitative Environmental Flow

The qualitative environmental flow of the Damodar River has been negatively affected by urban-industrial pollution and influent discharges containing various toxicants from mining areas and agricultural fields. The flow has also been altered

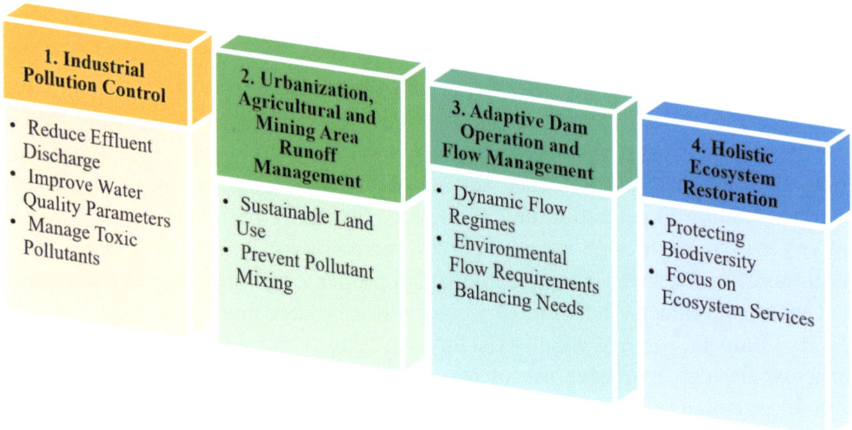

Fig. 4.7 Controlling factors for qualitative environmental flow improvement in the Damodar River

by the DVC dams and barrage (Hoque et al., 2022a; Chakraborty et al., 2021a). Urban and industrial growth, mining activities and intensive agricultural practice in the Damodar basin have contributed to water pollution, leading to reduced DO levels and increased BOD (Hoque et al., 2022a). The river's environmental flow quality can be improved by controlling industrial waste or effluent discharges to decrease toxic contamination and increase DO levels. This can also be done by implementing sustainable land management to prevent runoff pollution from urban, agricultural and mining areas. Additionally, regulating dam operations to allow more natural flow variability and releasing more water can help wash away pollutants. These efforts will promote good water quality and maintain a healthy river ecosystem, biodiversity and the overall health of the river environment, supporting the livelihoods of people who depend on healthy river ecosystems (Fig. 4.7).

A significant part of the river's water quality is poor due to the large amount of industrial effluents discharged into the river systems, especially downstream of the Durgapur barrage (Singh et al., 2022). A temporary improvement in water quality was observed as a result of industrial shutdowns during the COVID-19 pandemic and characterized by increased DO and BOD and decreased total dissolved solids (TDS) (Chakraborty et al., 2021a). This suggests that controlling industrial effluents will enhance water quality for aquatic biota (Hoque et al., 2025b). Besides, managing the discharge of toxic substances from the chemical industry is necessary to decrease overall contamination levels and generate more suitable habitats for aquatic life (Chakraborty et al., 2021a). Moreover, industrialisation, urbanization, and the expansion of modern agriculture with intensive mining activities lead to pollution in the Damodar River systems (Hoque et al., 2022a). Therefore, implementing better land management practices, including urban planning and sustainable agricultural practices, can prevent sewage and runoff from agricultural and mining areas from mixing with river water, thereby improving the quality (Hoque et al., 2022a; Chakraborty et al., 2021a; Jiang et al., 2022). Additionally, afforestation can prevent soil erosion and reduce sediment load and turbidity in the river systems (Buendia et al., 2016). The dams operated by the DVC have significantly changed the natural flow pattern of the river, impacting ecological stability (Hoque et al., 2022a). Hence, environmental flow management aims to ensure sufficient water provision to sustain and maintain aquatic ecosystems (Hoque et al., 2022a). Adjusting dam operations to release more water with natural flow variability rather than maintaining a constant minimum flow is critical to restoring river ecosystems and supporting biota (Hoque et al., 2022a). The current hazardous condition of the river, including the decline in fish habitats, indicates the need for a holistic or comprehensive approach to mitigate pollution and restore habitat (Wohl et al., 2015; Hoque et al., 2025a). Thus, improvement of water quality and restoration of river flow are crucial for preserving freshwater biodiversity and the ecosystem services they provide (Hoque et al., 2022a; Hoque & Islam, 2024). Apart from the qualitative environmental flow analysis, the dams play an important role in development. Thus, Chap. 5 delves into the appraisal of the development activities in the DRB about both the positive and negative aspects.

References

Aboyeji, O. O. (2013). Freshwater pollution in some Nigerian local communities, causes, consequences and probable solutions. *Academic Journal of Interdisciplinary Studies, 2*(13), 111. https://doi.org/10.5901/ajis.2013.v2n13p111

Adimalla, N. (2020). Controlling factors and mechanism of groundwater quality variation in semiarid region of South India: An approach of water quality index (WQI) and health risk assessment (HRA). *Environmental Geochemistry and Health, 42*(6), 1725–1752. https://doi.org/10.1007/s10653-019-00374-8

Ahmed, J., Wong, L. P., Chua, Y. P., & Channa, N. (2020). Drinking water quality mapping using water quality index and geospatial analysis in primary schools of Pakistan. *Water, 12*(12), 3382. https://doi.org/10.3390/w12123382

Al-Hadithi, M., Hasan, K., Algburi, A., & Al-Paruany, K. (2019). Groundwater quality assessment using irrigation water quality index and GIS in Baghdad, Iraq. *Jordan Journal of Earth and Environmental Sciences, 10*(1), 15–20.

Baron, J. S., Poff, N. L., Angermeier, P. L., Dahm, C. N., Gleick, P. H., Hairston, N. G., Jr., et al. (2002). Meeting ecological and societal needs for freshwater. *Ecological Applications, 12*(5), 1247–1260. https://doi.org/10.1890/1051-0761

Batarseh, M., Imreizeeq, E., Tilev, S., Al Alaween, M., Suleiman, W., Al Remeithi, A. M., ... & Al Alawneh, M. (2021). Assessment of groundwater quality for irrigation in the arid regions using irrigation water quality index (IWQI) and GIS-Zoning maps: Case study from Abu Dhabi Emirate, UAE. *Groundwater for Sustainable Development, 14*, 100611. https://doi.org/10.1016/j.gsd.2021.100611

Bazeli, J., Ghalehaskar, S., Morovati, M., Soleimani, H., Masoumi, S., Rahmani Sani, A., et al. (2020). Health risk assessment techniques to evaluate non-carcinogenic human health risk due to fluoride, nitrite and nitrate using Monte Carlo simulation and sensitivity analysis in groundwater of Khaf County, Iran. *International Journal of Environmental Analytical Chemistry, 102*(8), 1793–1813. https://doi.org/10.1080/03067319.2020.1743280

Bera, B., Bhattacharjee, S., Chamling, M., Ghosh, A., & Sengupta, N. (2021). Fluoride hazard and risk enumeration of hard rock unconfined aquifers in the extended part of Chhota Nagpur Gneissic Complex. *Journal of the Geological Society of India, 97*(2), 199–209. https://doi.org/10.1007/s12594-021-1651-0

Bhateria, R., & Jain, D. (2016). Water quality assessment of lake water: A review. *Sustainable Water Resources Management, 2*, 161–173. https://doi.org/10.1007/s40899-015-0014-7

BIS. (2012). *Water-specification*. I. S. D. Retrieved February 15, 2022, from http://cgwb.gov.in/Documents/WQ-standards.pdf

Brown, R. M., McClelland, N. I., Deininger, R. A., & O'Connor, M. F. (1972). A water quality index—crashing the psychological barrier. In *Indicators of environmental quality* (pp. 173–182). Springer. https://doi.org/10.1007/978-1-4684-2856-8_15

Buendia, C., Bussi, G., Tuset, J., Vericat, D., Sabater, S., Palau, A., & Batalla, R. J. (2016). Effects of afforestation on runoff and sediment load in an upland Mediterranean catchment. *Science of the Total Environment, 540*, 144–157. https://doi.org/10.1016/j.scitotenv.2015.07.005

Chakraborty, B., Roy, S., Bera, A., Adhikary, P. P., Bera, B., Sengupta, D., et al. (2021a). Cleaning the river Damodar (India): Impact of COVID-19 lockdown on water quality and future rejuvenation strategies. *Environment, Development and Sustainability, 23*(8), 11975–11989. https://doi.org/10.1007/s10668-020-01152-8

Chakraborty, B., Roy, S., Bera, A., Adhikary, P. P., Bera, B., Sengupta, D., et al. (2021b). Geospatial assessment of groundwater quality for drinking through water quality index and human health risk index in an upland area of Chota Nagpur Plateau of West Bengal, India. In *Spatial modeling and assessment of environmental contaminants* (pp. 327–358). Springer. https://doi.org/10.1007/978-3-030-63422-3_19

Chen, Y. (2021). Analysis on the evolution trend of seawater quality and eutrophication in Tianjin Hangu agricultural and fishery area from 2010 to 2019. In *E3S Web of Conferences* (Vol. 245, p. 02024). EDP Sciences. https://doi.org/10.1051/e3sconf/202124502024

Chen, J., He, D., & Cui, S. (2003). The response of river water quality and quantity to the development of irrigated agriculture in the last 4 decades in the Yellow River Basin, China. *Water Resources Research, 39*(3). https://doi.org/10.1029/2001WR001234

Citakoglu, H., Demir, A., & Gemici, B. (2021). Regionalization and mapping of dissolved oxygen concentration of Sakarya Basin by L-moments method. *Mühendislik Bilimleri ve Tasarım Dergisi, 9*(2), 495–510.

Deepika, S. S., & Singh, S. K. (2015). Water quality index assessment of Bhalswa lake, New Delhi. *International Journal of Advanced Research, 3*(5), 1052–1059.

Dhembare, A. J. (2012). Assessment of water quality indices for irrigation of Dynaneshwar dam water, Ahmednagar, Maharashtra, India. *Archives of Applied Science Research, 4*(1), 348–352.

Duvva, L. K., Panga, K. K., Dhakate, R., & Himabindu, V. (2022). Health risk assessment of nitrate and fluoride toxicity in groundwater contamination in the semi-arid area of Medchal, South India. *Applied Water Science, 12*(1), 1–21. https://doi.org/10.1007/s13201-021-01557-4

Eaton, F. M. (1950). Significance of carbonates in irrigation waters. *Soil Science, 69*(2), 123–134.

Eslamian, S., Parvizi, S., Ostad-Ali-Askari, K., & Hossein, T. (2018). *Encyclopedia of engineering geology: Water* (pp. 1–5). Springer. https://doi.org/10.1007/978-3-319-12127-7_295-1

Falowo, O. O., Akindureni, Y., & Ojo, O. (2017). Irrigation and drinking water quality index determination for groundwater quality evaluation in Akoko Northwest and Northeast areas of Ondo State, Southwestern Nigeria. *American Journal of Water Science and Engineering, 3*(5), 50–60. https://doi.org/10.11648/j.ajwse.20170305.11

Ghosh, S., Hoque, M. M., Saha, U. D., & Islam, A. (2022). Assessment of dam-induced changes in ecogeomorphological behaviour and fluvial functionality in the Damodar River, West Bengal, India. AQUA—Water Infrastructure, *Ecosystems and Society, 71*(6), 722–750. https://doi.org/10.2166/aqua.2022.003

Ha, N. N., Huong, T. T. T., & Van, T. T. (2021). Surface water pollution risk from Vietnam Water Quality Index (VN-WQI) in the Ca Mau City, Mekong Delta. *Nature Environment & Pollution Technology, 20*(4). https://doi.org/10.46488/NEPT.2021.v20i04.007

Hoque, M. M., Islam, A., & Ghosh, S. (2022a). Environmental flow in the context of dams and development with special reference to the Damodar Valley Project, India: A review. *Sustainable Water Resources Management, 8*(3), 1–27. https://doi.org/10.1007/s40899-022-00646-9

Hoque, M. M., Islam, A., & Mahammad, S. (2022b). Assessing the surface and bottom river water quality for drinking purpose and human health risk analysis: A study of Damodar River, India. *Arabian Journal of Geosciences, 15*(23), 1734. https://doi.org/10.1007/s12517-022-10987-6

Hoque, M. M., Islam, A., Sarkar, B., & Saha, U. D. (2024). Assessing the surface and bottom river water quality for irrigation: A study of Damodar River, India. *International Journal of Energy and Water Resources*, 1–18. https://doi.org/10.1007/s42108-022-00206-z

Hoque, M. M., Islam, A., Islam, A. R. M. T., Pal, S. C., Mahammad, S., & Alam, E. (2023a). Assessment of soil heavy metal pollution and associated ecological risk of agriculture dominated mid-channel bars in a subtropical river basin. *Scientific Reports, 13*(1), 11104. https://doi.org/10.1038/s41598-023-38058-0

Hoque, M. M., Islam, A., Islam, A. R. M. T., Das, B. C., Pal, S. C., Arabameri, A., & Khan, R. (2023b). Spatio-temporal assessment of water quality of a tropical decaying river in India for drinking purposes and human health risk characterization. *Environmental Science and Pollution Research, 30*(45), 101653–101668. https://doi.org/10.1007/s11356-023-29431-7

Hoque, M. M., Islam, A., Pal, S. C., Sarkar, B., & Hossain, Z. M. (2025a). Assessment of fish habitat suitability in the wake of elevated pollution of a tropical river in India. *Environmental Science and Pollution Research*, 1–15. https://doi.org/10.1007/s11356-024-35685-6

Hoque, M. M., Pal, S. C., Mandal, S., & Islam, A. (2025b). Assessment of mollusc habitat suitability of an anthropogenically altered tropical river in India. *Environmental Monitoring and Assessment, 197*(10), 1–18. https://doi.org/10.1007/s10661-025-14593-3

Hoque, M. M., & Islam, A. (2024). Spatio-temporal dynamics of ecological, bacteriological, and overall water quality of the Damodar River, India. *Environmental Science and Pollution Research, 31*(12), 18465–18484. https://doi.org/10.1007/s11356-024-32185-5

Inyinbor Adejumoke, A., Adebesin Babatunde, O., Oluyori Abimbola, P., Adelani Akande Tabitha, A., Dada Adewumi, O., & Oreofe Toyin, A. (2018). Water pollution: Effects, prevention, and climatic impact. In *Water challenges of an urbanizing world* (pp. 33–47). https://doi.org/10.5772/intechopen.72018

Islam, A., Sardar, N., Mohinuddin, S., Hoque, M. M., Sengupta, S., Das, B. C., ... & Sengupta, B. (2023). Quasi-equilibrium channel metamorphosis in planform of a subtropical river in India in post-dam period. *Catena, 221*, 106793. https://doi.org/10.1016/j.catena.2022.106793

Islam, A., Hoque, M. M., Ghosh, S., Sarkar, B., Mahammad, S., Das, B. C., ... & Alam, E. (2024). Hydro-chemical characterization and irrigation suitability assessment of a tropical decaying river in India. *Scientific Reports, 14*(1), 20096. https://doi.org/10.1038/s41598-024-70851-3

Islam, A., Das, B. C., Mahammad, S., Hoque, M. M., Pham, Q. B., Sarkar, B., ... & Barman, S. D. (2025). Assessing river water quality for ecological risk in the context of a decaying river in India. *Environmental Science and Pollution Research, 32*(16), 10223–10239. https://doi.org/10.1007/s11356-024-33684-1

Jannat, J. N., Khan, M. S. I., Islam, H. T., Islam, M. S., Khan, R., Siddique, M. A. B., et al. (2022). Hydro-chemical assessment of fluoride and nitrate in groundwater from east and west coasts of Bangladesh and India. *Journal of Cleaner Production, 372*, 133675. https://doi.org/10.1016/j.jclepro.2022.133675

Jayaswal, K., Sahu, V., & Gurjar, B. R. (2018). Water pollution, human health and remediation. In *Water remediation* (pp. 11–27). https://doi.org/10.1007/978-981-10-7551-3_2

Jiang, J., Zhao, G., Wang, D., Liu, L., Yan, X., & Song, H. (2022). Identifying trends and driving factors of spatio-temporal water quality variation in Guanting Reservoir Basin, North China. *Environmental Science and Pollution Research, 29*(58), 88347–88358. https://doi.org/10.1007/s11356-022-21714-9

Karataş, A., & Karataş, E. (2016). Environmental education as a solution tool for the prevention of water pollution. *Journal of Survey in Fisheries Sciences*, 61–70. https://doi.org/10.17762/sfs.v3i1.100

Kelley, W. P. (1963). Use of saline irrigation water. *Soil Science, 95*(6), 385–391.
Kumar, P., Srivastava, S., Banerjee, A., & Banerjee, S. (2022). Prevalence and predictors of water-borne diseases among elderly people in India: Evidence from Longitudinal Ageing Study in India, 2017–18. *BMC Public Health, 22*(1), 993. https://doi.org/10.1186/s12889-022-13376-6
Leal, M. F., de Simone, L. R. L., Lacerda, A. C. F., da Silva, E. L., & Pinheiro, T. G. (2021). Current distribution of the invasive mollusk Corbicula fluminea (OF Müller, 1774)(Bivalvia, Cyrenidae) in Brazil, including a new record from the state of Piauí. *Check List, 17*(1), 151–157. https://doi.org/10.15560/17.1.151
Li, H., Guo, S., Shin, K., Wong, M. S., & Henkelman, G. (2019). Design of a Pd–Au nitrite reduction catalyst by identifying and optimizing active ensembles. *ACS Catalysis, 9*(9), 7957–7966. https://doi.org/10.1021/acscatal.9b02182
Mahammad, S., & Islam, A. (2021). Evaluating the groundwater quality of Damodar Fan Delta (India) using fuzzy-AHP MCDM technique. *Applied Water Science, 11*(7), 1–17. https://doi.org/10.1007/s13201-021-01408-2
Maleki, A., & Jari, H. (2021). Evaluation of drinking water quality and non-carcinogenic and carcinogenic risk assessment of heavy metals in rural areas of Kurdistan, Iran. *Environmental Technology & Innovation, 23*, 101668. https://doi.org/10.1016/j.eti.2021.101668
Meinzen-Dick, R., & Bakker, M. (2001). Water rights and multiple water uses–framework and application to Kirindi Oya Irrigation System Sri Lanka. *Irrigation and Drainage Systems, 15*(2), 129–148. https://doi.org/10.1023/A:1012903127115
Mezbour, R., Reggam, A., Maazi, M. C., & Houhamdi, M. (2018). Evaluation of organic pollution index and the bacteriological quality of the water of the Lake of birds (ELTarf East-Algerian). *Journal of Materials and Environmental Sciences, 9*(3), 971–979. https://doi.org/10.26872/jmes.2018.9.3.108
Ostad-Ali-Askari, K., & Shayannejad, M. (2021). Quantity and quality modelling of groundwater to manage water resources in Isfahan-Borkhar Aquifer. *Environment, Development and Sustainability, 23*(11), 15943–15959. https://doi.org/10.1007/s10668-021-01323-1
Ostad-Ali-Askari, K., Shayannejad, M., & Ghorbanizadeh-Kharazi, H. (2017). Artificial neural network for modeling nitrate pollution of groundwater in marginal area of Zayandeh-rood River, Isfahan, Iran. *KSCE Journal of Civil Engineering, 21*, 134–140.
Picetti, R., Deeney, M., Pastorino, S., Miller, M. R., Shah, A., Leon, D. A., et al. (2022). Nitrate and nitrite contamination in drinking water and cancer risk: A systematic review with meta-analysis. *Environmental Research, 210*, 112988. https://doi.org/10.1016/j.envres.2022.112988
Raghunath, H. M. (1987). *Ground water: Hydrogeology, ground water survey and pumping tests, rural water supply and irrigation systems*. New Age International.
Richards, L. A. (1954). Diagnosis and improvement of saline and alkali soils. USDA hand book. Accessed 12 Dec 2021. https://www.ars.usda.gov/ARSUserFiles/20360500/hb60_pdf/hb60complete.pdf
Sarkar, B., & Islam, A. (2019). Assessing the suitability of water for irrigation using major physical parameters and ion chemistry: A study of the Churni River, India. *Arabian Journal of Geosciences, 12*(20), 1–16. https://doi.org/10.1007/s12517-019-4827-9
Sarkar, B., & Islam, A. (2021). Assessing the suitability of groundwater for irrigation in the light of natural forcing and anthropogenic influx: A study in the Gangetic West Bengal, India. *Environmental Earth Sciences, 80*(24), 1–19. https://doi.org/10.1007/s12665-021-10087-w
Sarkar, B., Islam, A., & Das, B. C. (2021). Role of declining discharge and water pollution on habitat suitability of fish community in the Mathabhanga-Churni River, India. *Journal of Cleaner Production, 326*, 129426. https://doi.org/10.1016/j.jclepro.2021.129426
Sarkar, B., Islam, A., Das, B. C., & Nandy, S. (2022). Corrosion and scaling potential of groundwater in Quaternary aquifers of Bengal Basin, India. *Arabian Journal of Geosciences, 15*(12), 1–21. https://doi.org/10.1007/s12517-022-10415-9
Seal, K., Chaudhuri, H., Pal, S., Srivastava, R. R., & Soldatova, E. (2022). A study on water pollution scenario of the Damodar river basin, India: Assessment of potential health risk using

long term database (1980–2019) and statistical analysis. *Environmental Science and Pollution Research, 29*(35), 53320–53352. https://doi.org/10.1007/s11356-022-19402-9
Shukla, S., & Saxena, A. (2020). Sources and leaching of nitrate contamination in groundwater. *Current Science, 118*(6).
Singh, A., Deo, B., & Singh, S. P. (2014). Risk analysis on the use of Damodar river water for drinking purposes. *International Journal of Current Engineering and Technology, 4*(1), 405–410.
Singh, R. K., Chaturvedi, A., & Kumari, K. (2019). Water-quality assessment of Damodar River and its tributaries and subtributaries in Dhanbad Coal mining areas of India based on WQI. *Sustainable Water Resources Management, 5*(2), 381–386. https://doi.org/10.1007/s40899-017-0159-7
Singh, G., Jindal, T., Patel, N., & Dubey, S. K. (2022). A coherent review on approaches, causes and sources of river water pollution: An Indian perspective. In *Soil-water, agriculture, and climate change: Exploring linkages* (pp. 247–271). Springer. https://doi.org/10.1007/978-3-031-12059-6_13
Son, C. T., Giang, N. T. H., Thao, T. P., Nui, N. H., Lam, N. T., & Cong, V. H. (2020). Assessment of Cau River water quality assessment using a combination of water quality and pollution indices. *Journal of Water Supply: Research and Technology-Aqua, 69*(2), 160–172. https://doi.org/10.2166/aqua.2020.122
Spandana, M. P., Suresh, K. R., & Prathima, B. (2013). Developing an irrigation water quality index for the Vrishabavathi command area. *International Journal of Engineering Research & Technology, 2*, 821–830.
Sunitha, V., Reddy, Y. S., Suvarna, B., & Reddy, B. M. (2021). Human health risk assessment (HHRA) of fluoride and nitrate using pollution index of groundwater (PIG) in and around hard rock terrain of Cuddapah, AP South India. *Environmental Chemistry and Ecotoxicology, 4*, 113–123. https://doi.org/10.1016/j.enceco.2021.12.002
Tiri, A., Belkhiri, L., Asma, M., & Mouni, L. (2020). Suitability and assessment of surface water for irrigation purpose. In *Water chemistry* (pp. 1–18). https://doi.org/10.5772/intechopen.86651
Todd, D. K. (1960). Salt water intrusion of coastal aquifers in the United States. *Subterranean Water, 52*, 452–461. https://doi.org/10.1016/j.ecolecon.2009.04.019
Ward, M. H., DeKok, T. M., Levallois, P., Brender, J., Gulis, G., Nolan, B. T., & VanDerslice, J. (2005). Workgroup report: Drinking-water nitrate and health—recent findings and research needs. *Environmental Health Perspectives, 113*(11), 1607–1614. https://doi.org/10.1289/ehp.8043
WBPCB. (2022). *Water quality: West Bengal Pollution Control Board water quality information system*. https://www.wbpcb.gov.in/
Wohl, E., Lane, S. N., & Wilcox, A. C. (2015). The science and practice of river restoration. *Water Resources Research, 51*(8), 5974–5997. https://doi.org/10.1002/2014WR016874
Yıldız, S., & Karakuş, C. B. (2020). Estimation of irrigation water quality index with development of an optimum model: A case study. *Environment, Development and Sustainability, 22*(5), 4771–4786. https://doi.org/10.1007/s10668-019-00405-5
Zhang, Y., Wu, J., & Xu, B. (2018). Human health risk assessment of groundwater nitrogen pollution in Jinghui canal irrigation area of the loess region, northwest China. *Environmental Earth Sciences, 77*(7), 1–12. https://doi.org/10.1007/s12665-018-7456-9

Chapter 5
Dams and Development

5.1 Dams, Economic Development in the Global Context

One of the most controversial issues in dam construction is that it promotes development on the one hand and degrades the ecosystems on the other hand (Hoque et al., 2022a). Dams are installed for sustainable water resource management based on the principle of economically viable, socially equitable, and environmentally sustainable (WCD, 2000). Dams induce development by supplying water and hydropower in agricultural fields, industries, and domestic sectors (Altinbilek, 2002; Loomis, 2002). Moreover, it creates potentiality for development by serving beneficiaries such as flood control, flow regulation, irrigation, water usage, hydropower generation, drought mitigation, recreation, fisheries, and navigation (WCD, 2000; Ghosh & Mistri, 2013; Tortajada, 2014).

Though a dam reservoir provides several benefits to improve human lives and livelihood, it also has several negative environmental and social impacts (Richter & Thomas, 2007a, 2007b). The most important social consequences are displacement and resettlement (Tortajada, 2014). About 40–80 million people have been displaced. At the global scale, loss of lives and property damage have been reported, especially for the downstream areas in the wake of dam failure. Dam construction is the main factor for physical degradation, fragmentation, and diversion of the aquatic and terrestrial watershed ecosystem (WCD, 2000), which affects human well-being, aquatic life, and the environment due to decreasing ecosystem services in the river, wetlands, and floodplains (Overton et al., 2014).

Against this backdrop, the present chapter explores three Damodar Valley Corporation (DVC) facets. The first segment (Sect. 5.2) focuses on the DVC's impact on agriculture, specifically addressing crop diversity, crop intensity, per capita agricultural land, cropping patterns, irrigation, and overall agricultural development. The second section (Sect. 5.3) delves into the local community's perspective, providing support for both the DVC's authority's assertions and the community's

M. M. Hoque, A. Islam, *Dams, Development and Environmental Flow in the Damodar River Basin*, https://doi.org/10.1007/978-3-032-11161-6_5

viewpoint regarding other developmental endeavours like transportation and embankment projects. The discourse encompasses topics such as soil erosion prevention, flood control, electricity generation and distribution, domestic and industrial water supply, establishment of fishing grounds, and the creation of recreational areas.

5.2 Damodar Valley Corporation (DVC) and Agriculture

The DVC reservoirs play a great role in irrigation facilities through water resource management in the middle and lower Damodar Valley. The total irrigation command area of the Damodar Valley Project (DVP) spans an area of 569,000 ha (5690 km^2). In the basin region, it generated 486,040 ha (4860.4 km^2) of irrigation potential land as of 2022, a significant increase from the 364,000 ha (3640 km^2) recorded in 2015–2016 (DVC, 2016). Additionally, it has created 463,782 ha (4637.82 km^2) of irrigation potential land for *Kharif* crops and 22,258 ha (222.58 km^2) for *Rabi* crops (DVC, 2022). Moreover, DVC has created 393,763 and 22,258 ha of irrigation potential for the *Kharif* and *Rabi* crops in West Bengal, while 67,542 ha for *Kharif* crops in Jharkhand (DVC, 2020). About 30,000 ha (300 km^2) of land in the upper valley is being irrigated, and every year, the irrigation potential land is increased for the availability of water from 17,172 check dams constructed by DVC (DVC, 2022). The construction of check dams in the basin area for creating irrigation potential leads to drastic changes in the agricultural land use, land pattern, and also cropping pattern in this region. After the dam's construction, vast areas of the fallow and wasteland became irrigable land in this region. Now in the basin area, large quantities of paddy and vegetables are produced due to the availability of water, which is provided by dams. Based on the Damodar Valley Reservoir Regulation Committee (DVRRC), water resource is allocated for *Kharif* and *Rabi* cultivation but not for *Boro* cultivation in the lower Damodar Valley region according to the Damodar River regulation. Water provided for *Kharif* crop irrigation during 2015–2016 was about 850,000 acre-ft and 70,000 acre-ft for *Rabi* crops (DVC, 2016). In 1955, the Durgapur barrage was constructed on the river Damodar to supply water for irrigation purposes in the districts of Bardhaman, Bankura, and Hooghly of West Bengal. This basin scale scenario indicates agricultural development in the post-dam period. The present work is intended to examine the village-level development of agriculture through an intensive field survey as discussed in the following sections.

The DVC dams have been installed in the basin area, and DVC manifests massive socio-economic benefits. In this context, the household survey is conducted to recognize the role of DVC dams in socio-economic development as well as the environmental deterioration in the study area in the pre-dam (1957) and post-dam (2022) periods. It is pertinent to mention here that 1957 was chosen as the pre-dam representative timeline based on the constitution year (1957) of 'The Lower Damodar Investigation Committee' to monitor the DVC works and their impact on

Table 5.1 Sample household from each village

Village name	Total households (2011)	The proportion of households affected by DVC dams (p)	Sample size
Urwan	776	0.895	31
Urro	232	0.825	34
Landhadi	51	0.801	30
Lalpania (CT)	723	0.881	35
Khetko	1510	0.775	75
Lohpiti	634	0.705	97
Sheopur	163	0.875	32
Hijuli (CT)	1498	0.69	113
Arddhagram	254	0.701	80
Pratappur	161	0.822	43
Randiha/ Rondia	459	0.815	55
Rajarampur	444	0.795	63

Source: Computed from the Census of India, 2011, which was the latest census when the survey was conducted

hydrogeomorphological and ecological changes (Bhattacharyya, 2011). Similarly, 2022 is considered the post-dam representative timeline when the field survey was actually conducted to assess the dam-induced socio-economic and environmental changes. A total of 12 villages have been selected purposively, keeping the spatial coverage of the basin area in mind. A total of 688 households have also been selected randomly from 12 villages of the basin area for assessing socio-economic development induced by the DVC dams (Table 5.1). Moreover, the data from annual reports of DVC from 2011 to 2022 (Govt. of India) have also been used for the present study. In the present investigation, crop diversity index, crop intensity, and per capita agricultural land have been measured to evaluate the agricultural development made by the DVC.

5.2.1 *Agricultural Development*

Agricultural advancement signifies the growth of the agricultural sector within a specific region or country. The progress in the agricultural sector leads to an increase in agricultural production, ensuring food security and an improved quality of life. According to the primary survey, 75.09% of respondents acknowledged that agricultural progress has been achieved in the study villages. Furthermore, in Jharkhand, 73.08% of surveyed individuals, and in West Bengal, 77.90% of respondents, agreed that agricultural progress has been observed in their respective localities. Additionally, all respondents from Urro and Sheopur villages in Jharkhand, as well as Pratappur, Randiha, and Rajarampur villages in West Bengal (Fig. 5.1), unanimously affirmed the presence of agricultural progress in their areas. Conversely,

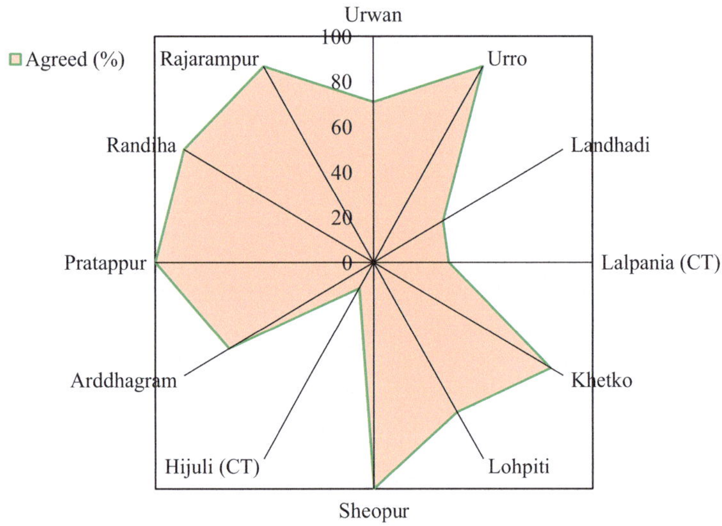

Fig. 5.1 Villagewise local people's perception regarding agricultural progress made by the Damodar Valley Corporation dams

86.73%, 65.71%, and 63.33% of surveyed individuals from Hijuli (CT), Lalpania (CT), and Landhadi villages, respectively, expressed disagreement, stating that no agricultural progress has taken place in their localities. Consequently, these findings suggest that, with a few exceptions, during the post-dam period, overall agricultural progress has been evident.

Following the construction of the dam, the agricultural sector experienced growth attributed to the development of irrigation facilities, increased usage of inorganic fertilizers and pesticides, and the integration of modern technology. As illustrated in Fig. 5.2a, b, there is a clear indication that the agricultural sector, along with its associated factors, progressed after the dam's establishment. About 56% (total sample size, i.e., $N = 688$) of respondents asserted that the agricultural sector's advancement took place in the post-dam period, while only 23.71% ($N = 688$) believed it occurred before the dam. Additionally, 1.94%–14.27% of respondents attributed the development of related factors to the pre-dam period. The consensus among respondents regarding agricultural development was notably high in the post-dam period. Interestingly, the ratings for agricultural development were relatively consistent between the pre- and post-dam periods, with respondents expressing mixed reactions in their assessments. It is worth noting that none of the respondents from Sheopur and Randiha villages acknowledged agricultural development in the pre-dam period. Similarly, no respondents from Sheopur village acknowledged it in the post-dam period. On the other hand, all respondents from Pratappur and Randiha villages unanimously agreed that agricultural development occurred in the post-dam era.

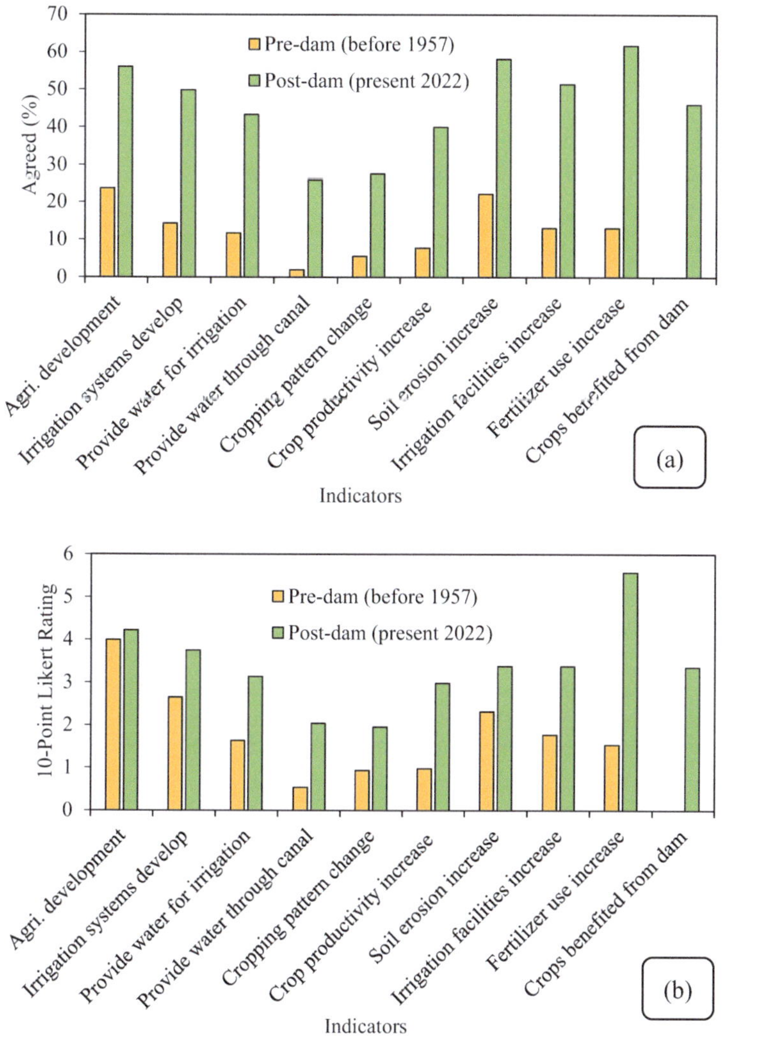

Fig. 5.2 Local people's perception regarding agricultural development and related to other factors before and after the DVC dam; (**a**) agreed and (**b**) rating

5.2.2 Development of Irrigation Facilities

Based on the perception survey, on average, 51.49% of respondents said that irrigation facilities have increased in the post-dam period (Fig. 5.2a). Besides, all the respondents of the Sheopur and Randiha villages disagreed with it in the pre-dam, whereas all the respondents of all the villages agreed with it in the post-dam. In addition, Rajarampur (42.86%) in the pre-dam period agreed with it, while a higher percentage of respondents in Sheopur (96.88%) and Rajarampur (96.83%) agreed with it in the post-dam period. Moreover, the rating for increasing irrigation facilities is given 1.77 out of 10 for the pre-dam and 3.38 for the post-dam period (Fig. 5.2b).

According to the perception survey, 49.59% of respondents have said that irrigation systems were developed in the post-dam period (Fig. 5.2a). Besides, no respondents from the Landhadi, Sheopur, and Randiha villages agreed that irrigation systems were developed in the pre-dam. Similarly, no respondents from Sheopur village agreed to it in the post-dam. However, 100% of respondents from Pratappur and Randiha villages agreed that it occurred in the post-dam. Besides, the rating for the development of irrigation systems is given as 2.65 out of 10 for the pre-dam and 3.75 for the post-dam period (Fig. 5.2b).

Based on the perception survey, 43.45% of respondents have said that water is being provided for irrigation in the post-dam period (Fig. 5.2a). No respondents of Urro, Landhadi, Sheopur, and Randiha villages agreed that water was being provided for irrigation in the pre-dam period. Likewise, none of the respondents of Sheopur village agreed to the post-dam irrigation facility. Besides, 100% of respondents of the Pratappur and Randiha villages confirmed that irrigation opportunities occurred in the post-dam period. The rating for providing irrigation water is 1.65 out of 10 for pre-dam and 3.13 for the post-dam period (Fig. 5.2b).

According to local perceptions, on average (average of the 12 study villages), 25.91% of respondents reported that water was supplied through the canal for irrigation in the post-dam period, whereas only 1.94% of respondents mentioned it for the pre-dam era (Fig. 5.2a). Specifically, no respondents from Urwan, Urro, Khetko, Sheopur, Hijuli (census town abbreviated as CT), Arddhagram, Randiha, and Rajarampur villages mentioned canal-based irrigation in the pre-dam period. However, respondents from Urro, Khetko, and Sheopur villages stated that water was being supplied through a canal in the post-dam period. Furthermore, 2.86%–3.33% of respondents from Landhadi, Lalpania (CT), and Lohpiti, along with 13.95% from Pratappur, reported canal-based irrigation in the pre-dam period. In contrast, a higher percentage of respondents from Randiha (86.27%) and Pratappur (69.77%) confirmed canal-based irrigation in the post-dam period. The rating for canal-based irrigation stands at 1.65 out of 10 for the pre-dam period, contrasting with a higher 3.13 rating for the post-dam phase (Fig. 5.2b).

In the study villages, 61.79% of respondents said that the use of fertilizer, including organic manures, has increased in the post-dam period (Fig. 5.2a). For example, 100% of respondents from Urwan, 96.83% from Rajarampur, and 83.64% from Randiha agreed to an increase in fertilizer use. Contrary to this, all the respondents of the Sheopur village disagreed with the increase in fertilizer use in the post-dam. The rating for fertilizer use of pre-dam is low (1.54 on 10 10-point Likert scale) compared to the post-dam period (5.57) (Fig. 5.2b).

In the present time, agricultural production depends on chemical fertilizers. However, this fertilizer is not suitable for the long-term production of any soil. That's why farmers use both organic and chemical fertilizers to maintain the fertility

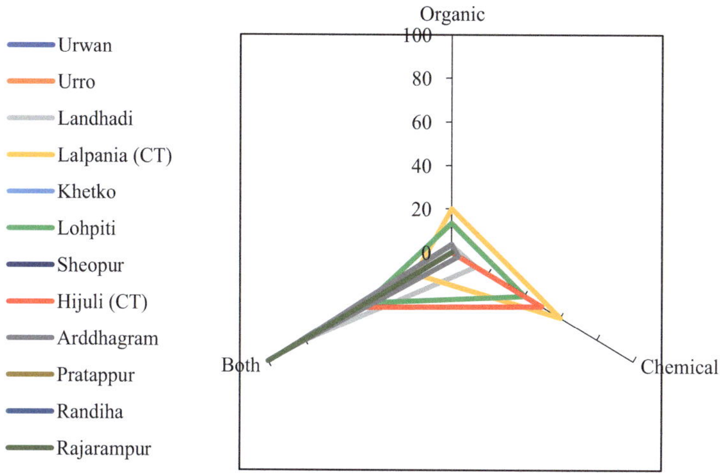

Fig. 5.3 Village-wise people's perception regarding fertilizer use

of the land. The 60% of Lalpania (CT), 49.56% of Hijuli (CT), 40.21% of Lohpiti, 13.33% of Landhadiand, and 3.75% of Arddhagram respondents said that only use chemical fertilizer in their irrigation fields. However, 20% of Lalpania (CT), 13.40% of Lohpiti, 3.75% of Arddhagram, and 3.33% of Landhadi people said that they use organic fertilizer in their agricultural fields (Fig. 5.3). Besides, 100% of Urwan, Urro, Khetko, Sheopur, Pratappur, Randiha, and Rajarampur said that they use both types of chemical fertilizer in their irrigation fields.

As per the perception survey, most of the villages use the dam or river water for irrigation. On average, 53.56% of respondents said that they used the river or dam water for irrigation purposes in the post-dam period, while 28% of respondents said they used it for irrigation in the pre-dam period. Only 14.29% of Lalpania (CT) and 16.81% of Hijuli (CT) respondents said that they use the river and dam water for irrigation, while a high 93.33% of Rajarampur, 83.64% of Randiha, and 82.35% of Urro respondents also said they use the river water for irrigation purposes in the post-dam period. Furthermore, the use of river water for irrigation was comparatively low in the pre-dam period (Fig. 5.4) as the irrigation equipment was not developed at that time and people depended on rainfall for their cultivation (Bhattacharyya & Wiley, 2014).

During the survey, it was observed that most farmers use dam or river water, but most farmers do not pay for it as they use it with their irrigation equipment. However, farmers of some villages like Urwan, Urro, Arddhagram, Pratappur, Randiha, and Rajarampur directly pay for the irrigation water (Fig. 5.5). Some people of Urwan and Randiha said they pay INR 300–400 per *Bigha* and Arddhagram and Rajarampur give INR 750–1200. These costs depend on the distance between the agricultural field and the location of the river or dam.

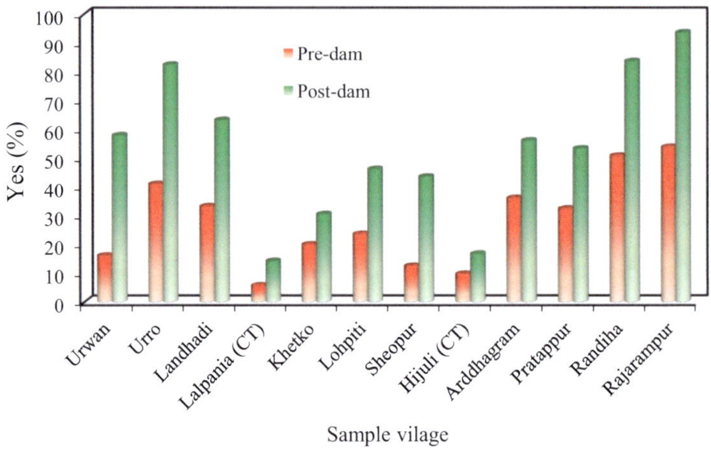

Fig. 5.4 Village-wise use of river water as a source of irrigation in the pre-and post-dam period

Fig. 5.5 Village-wise local people's perception regarding the cost of irrigation

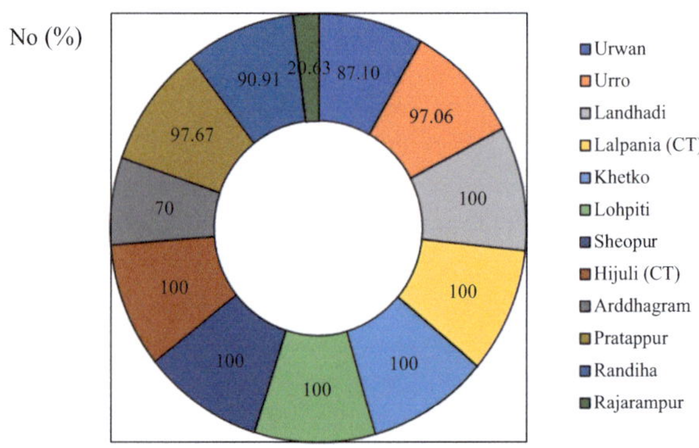

Through the comparative analysis of various factors related to agricultural development, it is observed that the agricultural condition was poor in the pre-dam periods (Fig. 5.2a, b). In this period, agriculture mainly depended on nature. Some respondents said that they faced 3–4 floods a year, which leads to huge economic loss. Transportation and communication at that time were not good. During the rainy season, everyone was stuck in their own house. Gathering food was the main challenge. After the dam, developing irrigation systems, creating canals, and controlling floods by dams, using fertilizer, pesticides, and modern technology made significant changes in the cropping pattern, cropping intensity, and overall agricultural practices, leading to a breakthrough in production. In post-dam periods, farmers produce various kinds of cash crops along with food crops. The dam creates a new route of transport for easily accessing the nearest town to sell their agricultural produce.

5.2.3 *Changing Cropping Pattern*

5.2.3.1 Crop Diversity

It is found that the crop diversity index ranges from 0.11 to 0.83 with a mean value of 0.37 in the pre-dam period, while it ranges from 0.20 to 0.86 with a mean value of 0.44 in the post-dam period (Table 5.2). It is found to be low for Hijuli (CT) village, while it is high for Urwan village in both (pre- and post-dam) periods. It is also found that the diversity of crops has positively changed from the pre-dam to the post-dam period. Higher crop diversity is observed for Lalpania (CT) and Khetko villages. Hence, the diversity of crops in the post-dam period is higher than in the pre-dam period. These results indicate that the cropping pattern has changed after DVC dams in the study villages, facilitating wider options of diversified crops for the farmers.

5.2.3.2 Cropping Intensity

During the pre-dam period, the cropping intensity was 100%. However, in the post-dam period, it ranged from 100.55% to 192.86%, with a mean value of 130.52% (Table 5.3). Notably, there was a spatial variation in cropping intensity among villages, with Lohpiti (100.55%) village showing lower values and Rajarampur village (192.86%) exhibiting higher values during the post-dam periods.

The analysis reveals a positive change in cropping intensity from the pre-dam to the post-dam period, with substantial increases observed in most villages, except for Urro, Landhadi, Lohpiti, and Sheopur. Overall, the cropping intensity during the

Table 5.2 Crop diversity (Gibbs-Martin index showing crop diversity of the selected villages in pre- and post-dam period)

	Crop diversity (Gibbs-Martin index)		
Village	Pre-dam (pre-1957)	Post-dam (2022)	Change
Urwan	0.83	0.86	0.03
Urro	0.45	0.47	0.02
Landhadi	0.34	0.42	0.08
Lalpania (CT)	0.43	0.54	0.11
Khetko	0.44	0.55	0.11
Lohpiti	0.41	0.48	0.07
Sheopur	0.43	0.50	0.07
Hijuli (CT)	0.11	0.20	0.09
Arddhagram	0.22	0.30	0.08
Pratappur	0.23	0.32	0.09
Randiha	0.25	0.30	0.05
Rajarampur	0.24	0.29	0.04

Computed by the authors, 2023

Table 5.3 Cropping intensity in the pre- and post-dam period of the selected villages

Village	Cropping intensity (%)		
	Pre-dam (pre-1957)	Post-dam (2022)	Change
Urwan	100	121	21
Urro		104.41	4.41
Landhadi		101.96	1.96
Lalpania (CT)		114.38	14.38
Khetko		116	16
Lohpiti		100.55	0.55
Sheopur		103.18	3.18
Hijuli (CT)		117.98	17.98
Arddhagram		147.88	47.88
Pratappur		161.90	61.90
Randiha		184.09	84.09
Rajarampur		192.86	92.86

Computed by the authors, 2023

post-dam period surpassed that of the pre-dam period. These results underscore a significant enhancement in the utilization of agricultural land, indicating a noteworthy increase in year-round crop production facilitated by irrigation from DVC dams in the study area.

Furthermore, as per the results of the perception survey, 27.60% of respondents indicated an increase in cropping intensity during the post-dam period, contrasting with only 5.55% of respondents who acknowledged such changes before the dam (Fig. 5.2a). Furthermore, during the pre-dam phase, all respondents from Urro, Lalpania (CT), Lohpiti, Sheopur, Randiha, and Rajarampur villages disagreed with the notion. However, in the post-dam period, respondents from Sheopur and Hijuli (CT) villages also expressed disagreement. In the pre-dam period, less than 3.33%–12.90% of respondents from the remaining villages agreed with the increased cropping intensity, but in the post-dam period, 73.02% of Rajarampur and 69.77% of Pratappur respondents acknowledged such changes. The rating assigned to the change in cropping intensity scored 0.93 out of 10 for the pre-dam period and 1.96 for the post-dam period (Fig. 5.2b), highlighting a minimal perceived impact on cropping intensity. Notably, the rating is significantly higher for the post-dam period.

5.2.3.3 Crop Productivity

Crop productivity of selected villages has also been investigated in pre- and post-dam periods**.** It is found that crop productivity has changed in the post-dam period. For example, the average rice production in the study area was 1.8 quintal/*Bigha* in the pre-dam period. It is 6.57 quintal/*Bigha* in the post-dam period. Besides, the average production of wheat, mustard, maize, gram, and tur (*arhar*) was 0.86, 0.45, 0.84, 0.38, and 0.30 quintal/*Bigha,* respectively, in the pre-dam period, while, it was

5.07, 1.24, 4.72, 1.02, and 0.54 quintal/*Bigha* for wheat, mustard, maize, gram, and tur (*arhar*) correspondingly in the post-dam period. Moreover, the village-wise distribution of change in crop production is given in Table 5.4.

Furthermore, according to the perception survey in the study villages, on average, 40.05% of respondents said that crop productivity has increased in the post-dam period, while only 7.80% of respondents agreed with it before the dam (Fig. 5.2a). Moreover, all the respondents of Landhadi, Lalpania (CT), Sheopur, Hijuli (CT), and Randiha villages disagreed with it in the pre-dam. Similarly, all the respondents of the Sheopur village disagreed with it in the post-dam. However, 3.09%–20% of respondents for the remaining village agreed with it for the pre-dam period, while a high percentage of Rajarampur (90.48%) and Urro (73.53%) respondents agreed with it for the post-dam. Besides, the rating for cropping pattern change is given 0.98 out of 10 for the pre-dam and 2.98 for the post-dam period (Fig. 5.2b).

Furthermore, on average, 46.06% of respondents said that crops have benefited from the dam as per the perception survey (Fig. 5.2a). Moreover, 100% of the respondents of Pratappur village said that crops benefited from the dam only. However, only 3.54% of the respondents of Hijuli (CT) agreed to it. In addition, 50–93.65% of Urwan, Sheopur, Arddhagram, and Rajarampur have said that crops benefited from dams. The average rating for it is 3.35 out of 10 (Fig. 5.2b).

5.2.4 *Dynamics in Agricultural Land and Land Use*

Total irrigated land was 750,860 ha in the pre-dam period, while it is 1,190,860 ha in the post-dam period in the basin area, resulting in a 58.60% increase in the irrigable land. Moreover, the availability of average irrigable agricultural land varies from 1.21 to 4.19 *Bigha* with a mean of 2.05 *Bigha* in the pre-dam while it varies from 1.46 to 3.04 *Bigha* with a mean of 2.20 *Bigha* in the post-dam period. Besides, in all the villages, except Urwan and Lalpania (CT), the irrigable agricultural land increased in the post-dam period (Table 5.5). The irrigable agricultural land of the Urwan and Lalpania (CT) villages decreases due to the submergence of irrigable agricultural land in the reservoir.

As per the local people's perception, the average availability of per capita agricultural land of the farmers is 14.47 *Katha* and ranges from 0.14 to 75 *Katha* in the study area. This is slightly higher than the Indian average (9.49 *Katha*) but lower than the world average (22.93 *Katha*) based on the Press Information Bureau, Government of India (2023). Moreover, the intra- and inter-village variability of per capita agricultural land availability is also observed. The higher variability of per capita agricultural land availability is observed in the villages of Urwan, Urro, Landhadi, Khetko, Lohpiti, and Rajarampur (Fig. 5.6). Besides, the average per capita agricultural land availability of the farmers is high at Lohpiti (22.67 *Katha*) and Randiha (22.81 *Katha*) villages, while it is low at Urwan (8.09 *Katha)* village (Fig. 5.6).

Table 5.4 Village-wise distribution of crop productivity (Quintal/*Bigha*) for the selected crop in the pre- and post-dam period

	Pre-dam (pre-1957)						Post-dam (2022)					
Village	Rice	Wheat	Mustard	Maize	Gram	Tur (*Arhar*)	Rice	Wheat	Mustard	Maize	Gram	Tur (*Arhar*)
Urwan	2.00	0.80	0.35	0.70	0.15	0.14	3.49	4.18	1.04	3.54	0.19	0.16
Urro	1.23	0.40	0.30	0.60	0.001	0.001	2.00	4.03	1.02	0.97	0.002	0.001
Landhadi	1.20	0.30	0.25	1.10	0.12	0.13	10.20	6.50	0.50	6.20	0.14	0.15
Lalpania	1.87	0.20	0.27	0.85	0.26	0.16	3.80	2.60	0.61	4.16	0.28	0.17
Khetko	1.70	0.21	0.24	0.82	0.24	0.15	3.90	2.50	0.60	4.10	0.27	0.16
Lohpiti	1.00	0.50	0.28	0.95	0.11	0.12	10.22	6.00	0.70	6.15	0.13	0.15
Sheopur	1.80	0.75	0.40	1.00	0.12	0.14	10.23	6.58	1.00	6.10	0.14	0.16
Hijuli	1.50	1.00	0.45	0.73	0.75	0.17	5.62	5.36	1.44	3.70	2.29	0.18
Arddhagram	2.00	1.50	0.60	0.84	0.62	0.64	6.93	5.50	1.86	4.85	1.85	1.32
Pratappur	2.10	1.00	0.55	0.75	0.65	0.60	6.97	5.58	1.80	4.88	1.88	1.30
Randiha	2.00	1.60	0.74	0.85	0.80	0.68	7.73	6.00	2.10	5.95	2.52	1.34
Rajarampur	3.20	2.00	1.00	0.90	0.73	0.70	7.78	6.06	2.16	5.97	2.51	1.35

Computed by the authors, 2023

Table 5.5 Village-wise distribution of average availability of irrigable agricultural land in the pre- and post-dam period

Village	Average availability of irrigable agricultural land (*Bigha*)		
	Pre-dam (pre-1957)	Post-dam (2022)	Change (%)
Urwan	2.22	2.10	−5.42
Urro	1.72	2.11	22.52
Landhadi	2.01	2.44	20.98
Lalpania (CT)	4.19	2.24	−46.51
Khetko	1.24	1.74	40.16
Lohpiti	2.13	2.97	39.51
Sheopur	1.21	1.46	20.74
Hijuli (CT)	1.60	2.00	25.00
Arddhagram	1.63	1.79	9.82
Pratappur	1.74	2.03	16.95
Randiha	2.66	3.04	14.47
Rajarampur	2.20	2.53	15.16

Computed by the authors based on the field survey, 2023

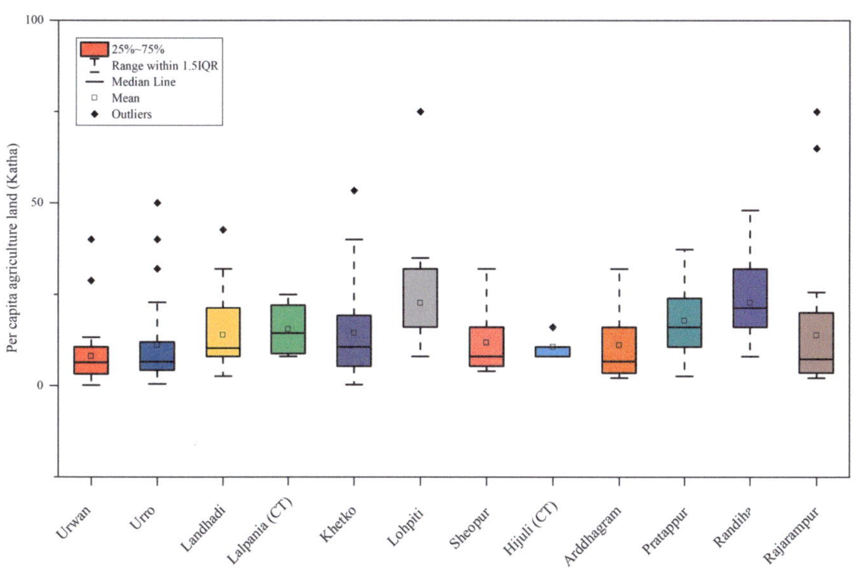

Fig. 5.6 Village-wise availability of per capita agricultural land

According to the local people's perception, on average, 54.72% of respondents with a rating of 6 (on a 10-point scale) said that the availability of agricultural land has changed in the last 20–30 years. Moreover, 74.29%–100% of surveyed people of Urwan, Lalpania (CT), Khetko, Arddhagram, and Rajarampur said that the agricultural land has changed in the last 20–30 years. Additionally, 95.35% of Pratappur, 93.75% of Sheopur, and 88.50% of Hijuli (CT) respondents have said that no change has occurred in agricultural land use in the past 20–30 years with regard to the survey year 2022.

According to the perception survey, Sheopur, Hijuli (CT), and Pratappur are three villages that recorded very little bit of agricultural land use change. Nearly 10% of agricultural land use has changed in the past 30 years (Fig. 5.7). However, Khetko is an exceptional village where 100% of surveyed people said that their agricultural land decreased due to dam construction. According to them, the reservoir water flows over their land area, and the total land area is submerged under water. Although 100% of % people of Rajarampur said that their agricultural land use changed, 53.97% of the people said their agricultural land use increased due to the development of the irrigation system, whereas the remaining percentage of people said their agricultural land decreased due to settlement growth. Moreover, 74.29% of the respondent of Lalpania (CT) and 64.52% of Urwan villages said that their agricultural land has been lost due to the creation of a dam. Besides, 47.50% of respondents of Arddhagram, 41.18% of Urro, and 40% of Randiha said that their agricultural land increased in the last 20–30 years (Fig. 5.7).

The agricultural lands of the villager lie within 2 km from their home, except for the Hijuli (CT) and Lohpiti villages (Fig. 5.8). The average distance between their house and agricultural land is 1.41 km. Moreover, the distance of agricultural land is 3.67 km for Hijuli (CT) and 2.22 km for Lohpiti villagers. Most of the villagers have an average of 1–2 *bighas* of agricultural land, and it is arable and fertile land. Besides, some respondents have said they also have a small amount of uncultivated land. Additionally, most of the villagers' agricultural land lies within 2–3 km of the river. The average distance between the agricultural land and the river is 1.51 km. Most of the villagers' agricultural land lies within 2 km, except Lalpania (2.5 km), Lohpiti (2.69 km), and Arddhagram (2.64 km) (Fig. 5.8).

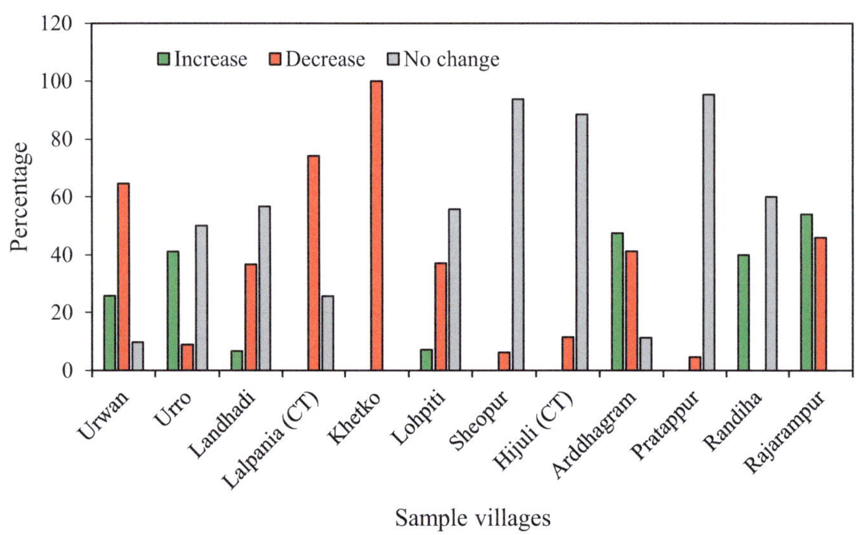

Fig. 5.7 Village-wise distribution of agricultural land changes in the last 20–30 years

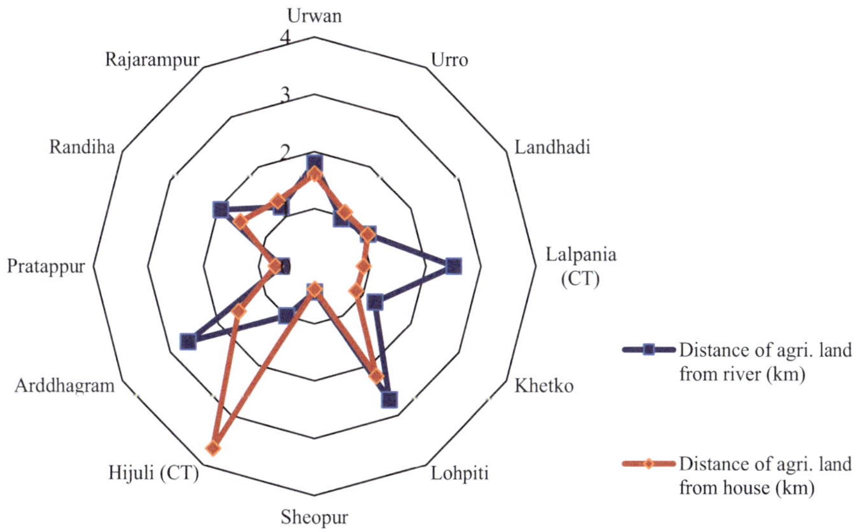

Fig. 5.8 Village-wise distance of agricultural land from the river and house

Table 5.6 One-way ANOVA showing the variations in agricultural development in the pre- and post-dam period

Component	*F*	*P*-value	*F*-critical	Remarks
Agricultural development based on local people's perception (% of agreed respondents)	46.021	4.39042E-06	4.494	Alternative
Agricultural development based on local people's perception (on a 10-point Likert scale)	9.423	0.007330623	4.494	Alternative
River water as a source of irrigation based on local people's perception (% of agreed respondents)	8.637	0.007596048	4.301	Alternative
Cropping intensity (index value)	10.325	0.004004491	4.301	Alternative
Crop diversity (index value)	0.933	0.344615844	4.301	Null
Crop productivity (quintal per bigha)	57.036	1.51518E-07	4.301	Alternative

Computed by the authors, 2023

5.2.5 *Pre-dam and Post-dam Variation in Agricultural Development*

To find out the statistically significant difference between the pre-dam and post-dam development in agriculture, one-way Analysis of Variance (ANOVA) is run using perceptual rating and indices (Table 5.6). The results show that five agricultural development parameters (overall agricultural development measured in % and rating scale, river water as a source of irrigation, cropping intensity, and cropping productivity) out of six exhibit statistically significant variations at a 0.05

significance level (Table 5.6). However, crop diversity does not reflect any significant difference in the pre-dam and post-dam conditions. This establishes that agrarian development has been recorded on a positive scale in the post-dam period.

5.3 Damodar Valley Corporation and Other Developmental Activities

The DVC is a multi-purpose river valley project with many facilities for controlling the colossal floods, developing irrigation systems, supplying domestic and industrial water, generating hydropower, checking soil erosion, boosting fishing, recreation tourism, and also increasing vegetation cover (Damodar Valley Corporation, 2021). Thus, it was once called the 'temple of modern India' by the then Prime Minister Pandit Jawaharlal Nehru (Khorakiwala, 2017). This section examines other development activities like afforestation, soil erosion check, flood control, and transport infrastructure development in the following sections.

5.3.1 Afforestation

The DVC, in collaboration with the forest department, Govt. of West Bengal, has planned for landscaping and beautification in the lower valley and created more than 7000 acres of land for a greenbelt (DVC, 2016). Moreover, DVC itself works in the upper valley for afforestation. A total of more than 2 lakh ha of afforestation land has been created in the upper valley (DVC, 2016). The greenbelt's developmental programme has been executed in the lower valley of Mejia and along the DVC canal. The DVC has also performed afforestation work on degraded forest land areas in the valley region.

So far as the perception of the local people is concerned, 46.82% of respondents agreed that the afforestation occurred in the study area. Moreover, no respondent was found for Lohpiti village, while 100% of the respondents of Sheopur village agreed to it. However, the rating for afforestation is given very low (the average, 2.87 out of 10) (Table 5.7).

5.3.2 Transport and Embankment

A total of 11.02 km of village roads has been constructed by DVC during 2021–2022 (DVC, 2022). The embankment from Dishergarh (Barakar and Damodar confluence point) towards Barddhaman, measuring 93 km, was constructed along the left bank of the Damodar River. In the post-dam period, the height of this embankment is increased downstream. The left side embankment's height around the Barddhaman

Table 5.7 Local people's perception regarding afforestation

	Affirmative response (%)		Rating (out of 10)	
Village	Average	SD	Average	SD
Urwan	16.13	42	1.22	2.39
Urro	32.35	48	2.03	2.98
Landhadi	40	51	2.82	2.62
Lalpania (CT)	37.14	49	2.73	3.69
Khetko	77.33	43	2.1	1.71
Lohpiti	0	0	0	0
Sheopur	100	0	4.73	2.16
Hijuli (CT)	33.63	49	1.93	2.76
Arddhagram	56.25	45	4.9	2.7
Pratappur	23.26	43	0.91	1.87
Randiha	72.73	43	5.6	2.65
Rajarampur	73.02	48	2.57	2.50

Computed by the authors, 2023

town was further increased, and the right bank embankment was destroyed after the devastating flood of 1885 (Ghosh & Guchhait, 2014). The initiative for the construction of an embankment around the Mithon reservoir has been taken to increase the storage capacity.

5.3.3 *Soil Erosion Check*

The soil conservation department of DVC has been implementing afforestation work in the upper valley, and it created 17,172 check dams for controlling soil erosion (DVC, 2022). It conserves 4 lakh ha of soil in the Damodar Valley. Moreover, according to the perception survey, on average, 58.12% of respondents have opined that soil erosion has increased in the post-dam period, while only 22.16% of respondents agreed with it before the dam (Fig. 5.2a). The rating for increasing soil erosion is given as 2.32 out of 10 for the pre-dam and 3.38 for the post-dam period (Fig. 5.2b). Additionally, all the respondents of the Sheopur and Randiha villages disagreed with increasing soil erosion in the pre-dam period. In addition, <40% of respondents for the remaining village of the pre-dam period agreed with it. A higher proportion of the respondents in Sheopur (96.88%), Randiha (70.91%), and Rajarampur (69.84%) agreed with it for the post-dam period.

5.3.4 *Flood Control*

Controlling floods is one of the major objectives of the DVC dams, as several devastating floods occurred in the valley region. The Valley region comes under the southwest monsoon of the Indian climate that hits during the period from the 4th week of July to the 1st week of August, causing heavy rainfall in the catchment area (Hoque et al., 2023; Islam et al., 2023). The normal average rainfall occurs at 1335 mm in the Damodar catchment area (Ghosh et al., 2016).

The original plan of DVC proposed seven reservoirs with a flood storage capacity of 2.915 million acre-feet in the Damodar Basin (DVC, 2022). In the first stage, four multi-purpose dams, Tilaiya (1953), Konar (1955), Maithon (1957), and Panchet (1959) were planned for a 1.51 million-acre feet flood water capacity reservoir. However, due to a lack of land acquisition, it gained 1.047 million-acre feet of flood reserve capacity (DVC, 2022). The creation of these reservoirs reduces flood probability in the Damodar Valley region by managing the Damodar River water during the monsoon period. The DVC dams have a 6.51 lakh cusec to 2.5 lakh cusecs flood moderation capability in the basin area.

Floods basically occur in the lower Damodar. Being the terminal reservoirs, Maithon and Panchet have a great role in controlling these floods in the lower Damodar. Maithon and Panchet reservoirs have a total of 542.76 MCM and 1086.76 MCM of floodwater reserve capacity (Fig. 5.9). Panchet has the highest floodwater reserve capacity (1086.76 MCM), and Konar has the lowest floodwater reserve capacity (55.51 MCM). The total flood reserve capacity under the four dams' reservoirs of DVC is 1862.66 MCM. The 63.74% of flood moderation was achieved by Maithon and Panchyet during 1958–2007 (Table 5.8).

Perception survey reveals that on average, 64.41% of respondents expressed that the flood probability in the basin area decreased after the DVC dams. Respondents gave the rating for it as low, with an average of 4.19. Besides, 100% of the respondents of the Sheopur and Pratappur villages mentioned that flood probability decreased after the dams, while only 10.31% of respondents of Lohpiti village agreed with it (Table 5.9). However, on average, 71.88% of respondents said that a dam is an effective tool for controlling floods, and the average rating for it was 4.99 (Table 5.9).

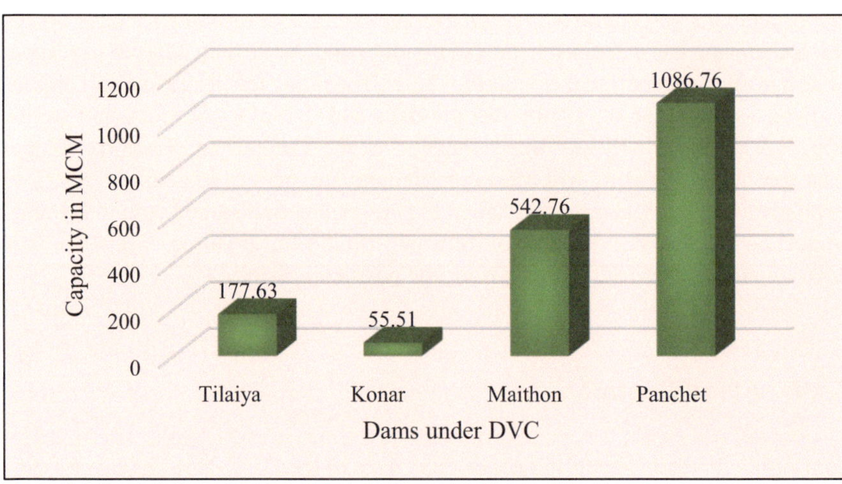

Fig. 5.9 Flood control capacity of DVC dams

Table 5.8 Flood moderation by Maithon and Panchet dams

Date	Max. inflow (m^3/s)	Max. outflow (m^3/s)	Flood moderation achieved (m^3/s)	Flood moderation achieved (%)
16–17 September, 1958	15,717.60	4956	10,761.00	68.46
1–2 October, 1959	17,646.36	8156.16	9487	53.76
27 September, 1960	9855.36	2605.44	7249.92	73.56
2–3 October, 1961	14,643.12	4559.52	10,053.6	68.66
2–3 October, 1963	12,772.32	3426.72	9345.6	73.17
24–25 October, 1963	13,168.8	2577.12	10,591.68	80.43
29–30 July, 1964	10,563.36	2208.96	8354.4	79.09
3–4 September, 1970	8269.44	2265.6	6003.84	72.60
16–18 July, 1971	12,007.68	5125.92	6881.76	57.31
12–13 October, 1973	16,652.16	4956	11,696.16	70.24
26–27 September, 1975	9742.08	3143.52	6598.56	67.73
27 September, 1978	21,919.68	4616.16	17,303.52	78.94
27–28 August, 1980	9657.12	4219.68	5437.44	56.30
25 June, 1984	10,600	4800	5800	54.72
14–17 September, 1993	7000	2800	4200	60.00
29 September, 1995	17,360.16	7080	10,280.16	59.22
25 September, 1999	10,300	3400	6900	66.99
19–21 August, 2000	10,800	5700	5300	49.07
24 September, 2006	14,400	6900	7500	52.08
25 September, 2007	11,100	7500	3600	32.43

Source: Based on Bhattacharyya (2011), Ghosh and Guchhait (2014)

5.3.5 *Electricity Generation and Provision*

The DVC has a total installed capacity of 174.2 megawatts (MW) of hydroelectricity generation from hydropower stations. The main hydel projects under the DVC are Maithon, Panchet, and Tilaiya hydel stations with a total power generation

Table 5.9 Local people's perception regarding flood probability and the stopping of flood

Village	Decreasing flood probability after dam		An effective method of stopping a flood	
	Yes	Rating (out of 10)	Yes	Rating (out of 10)
Urwan	54.84	4.7	67.74	5.91
Urro	73.53	4.62	52.94	3.76
Landhadi	36.67	3.84	43.33	3.12
Lalpania (CT)	37.14	2.13	51.43	3.4
Khetko	60	3.03	100	7.07
Lohpiti	10.31	0.6	26.80	2.03
Sheopur	100	4.8	100	4.87
Hijuli (CT)	69.91	3.96	66.37	4.41
Arddhagram	53.75	2.6	83.75	5.14
Pratappur	100	6.36	86.05	5.95
Randiha	80	6.16	87.27	6.52
Rajarampur	96.83	7.47	96.83	7.7

Computed by the authors, 2023

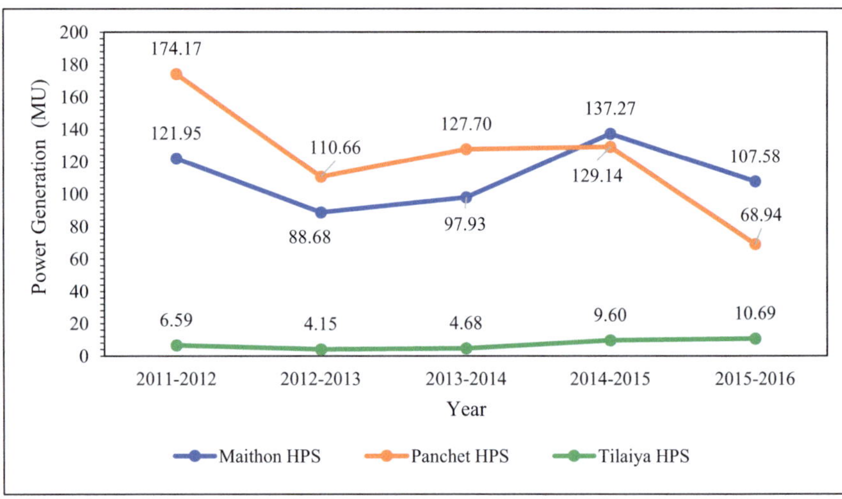

Fig. 5.10 Power generation by DVC hydel stations from 2011–2016. (Source: Annual report of DVC (2011–2016)

capacity are 63.2, 80, and 4 MW, respectively. Total hydroelectricity generation during 2015–2016 was 187.201 million units (MU) as against 276.004 MU in 2014–15. A decreasing trend is observed for the Maithon and Panchet hydel stations, while an increasing trend is observed for Tilaiya hydel station from 2011 to 2016 (Fig. 5.10).

Moreover, according to local people's perception, on average, 42.81% of respondents opined that the energy provision in the basin area improved after the DVC dams. Besides, 100% of the respondents of the Khetko and Sheopur villages

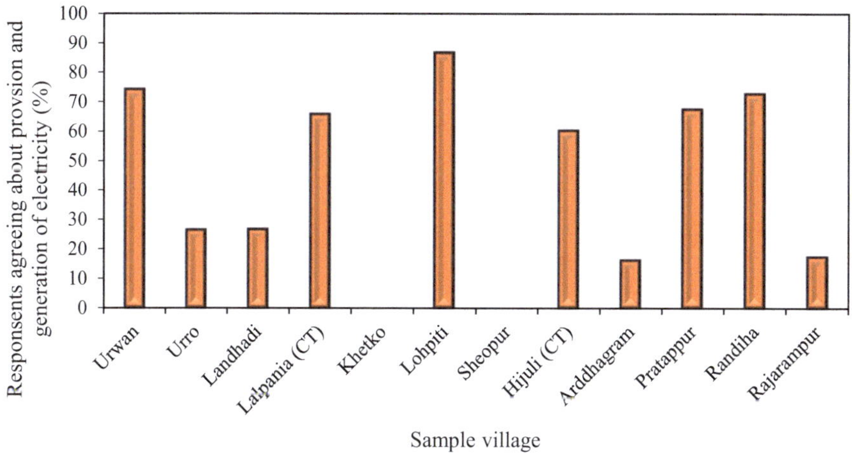

Fig. 5.11 Dam improved energy provision based on the local people's perception

disagreed with it (Fig. 5.11). A higher proportion of respondents (86.60%) from Lohpiti village mentioned that DVC dams improved the energy provision in their locality.

5.3.6 *Domestic and Industrial Water Supply*

With the availability of water resources and their proper management by DVC, many industries have grown in the Damodar Valley region ('*Ruhr*' of India) (Majumder et al., 2010). The DVC has a great role in coming up with these industries and their smooth growth in the region. The DVC provides water resources to 153 agencies of Jharkhand (86) and West Bengal (67) for domestic and industrial uses. Four DVC reservoirs store 419 million cubic metres of water out of the 680 cusecs water supply to meet industrial, municipal, and domestic requirements in West Bengal and Jharkhand. Every year, 340 million cubic metres of water are provided to municipal and industrial agencies within the valley region.

5.3.7 *Creation of a Fishing Ground*

One of the multi-purpose uses of DVC dams was to pursue the development of freshwater aquaculture in the reservoirs. This led to the establishment of the Maithon Fish Farm in 2008–2009. About 40.50 lakh fingerlings were produced in the Maithon Fish Farm, and they were released in the Maithon and Panchet reservoirs in the year 2012–2013 (DVC, 2013). Moreover, DVC created 3 fish landing centres

Table 5.10 Creation of reservoirs for fishing ground

	Affirmative response (%)		Rating (out of 10)	
Village	Average	SD	Average	SD
Urwan	100	0	8.56	0.92
Urro	91.18	26	5.9	2.47
Landhadi	70	21	6.5	1.10
Lalpania (CT)	62.86	48	4.67	3.75
Khetko	80	41	2.47	1.68
Lohpiti	56.70	44	2.81	2.75
Sheopur	100	0	4.83	1.56
Hijuli (CT)	50.44	51	2.89	3.41
Arddhagram	80	20	5.8	1.91
Pratappur	79.07	35	4.86	2.32
Randiha	100	0	8.19	1.24
Rajarampur	87.30	35	6.27	3.15

Computed by the authors, 2023

at Konar, 3 at Pacnchet, and 2 at Maithon reservoir from 2009–10 to 2011–2012. Further, 3 fish landing centers have also been created at Tilaiya Reservoir in 2012–13 (DVC, 2013).

On the perception survey, on average, 79.80% of respondents expressed the belief that establishing a reservoir is a favourable initiative for fishing grounds. The average rating assigned to this idea is 5.31. Besides, 100% of respondents from the Urwan, Sheopur, and Randiha villages unanimously agreed with this notion (Table 5.10).

5.3.8 *Recreation Area*

The DVC dams were created to pursue the development of tourism around the reservoir area as per their secondary objective. People use the reservoir for leisure activities, also as swimming, picnics, fishing, functions, etc. According to the primary survey, an average rating of 4.76 with 74.05% of respondents opining that the construction of dam reservoirs leads to the creation of a picnic ground around the reservoirs. People's perceptions regarding other recreational facilities created by the DVC dams are given in Table 5.11.

5.4 Dams and Displacement

One of the most significant social consequences of dam-reservoir creation is displacement, which has long-term negative impacts, including impoverishment, marginalization, and loss of livelihood. Marginalized people, especially rural

Table 5.11 Creation of recreation area by DVC dams

Village	Swimming		Picnic		Canoeing		Boating		Socio-cultural & related functions		Religious purpose	
	Yes	Rating	Yes	Rating	Yes	Rating	Yes	Rating	Yes	Rating	Yes	Rating
Urwan	54.84	3.6	83.87	7.84	0	0	80.65	6.08	64.52	4.68	25.81	2.83
Urro	20.59	0.86	41.18	2.62	0	0	14.71	0.39	8 82	0.71	2.94	0.31
Landhadi	0	0.18	63.33	4.5	0	0	6.67	0.36	0	0	0	0
Lalpania (CT)	94.29	6.15	100	7.37	22.86	0.89	22.86	1.11	34.29	1.93	25.71	1.19
Khetko	0	0	62.67	1.7	0	0	0	0	0	0	0	0
Lohpiti	29.90	2.71	62.89	4.38	6.19	0.48	6.19	0.48	6 19	0.48	13.40	1.14
Sheopur	6.25	0.4	100	2.9	0	0	0	0	0	0	0	0
Hijuli (CT)	83.19	5.33	100	7.33	13.27	0.7	13.27	0.78	23.01	1.74	7.08	0.22
Arddhagram	0	0	56.25	3.92	3.75	0	13.75	0.83	15.25	1.13	0	0
Pratappur	67.44	3.14	74.42	4.23	23.26	0.86	51.16	2.77	15.28	0.18	23.26	0.32
Randiha	47.27	2.89	83.64	6.15	100	0	7.27	0.37	0	0	0	0
Rajarampur	36.51	2.07	60.32	4.17	9.52	0.2	36.51	2.07	39.68	2.1	20.63	1.57

Computed by the authors, 2023; Note: 10-Point Likert Rating

communities, are the most affected by displacement. Displacement impacts vulnerable groups like tribal people, women, and children. Displacement leads to negative impacts, including poverty, joblessness, food insecurity, poor health, community resources loss, and psychological stress. The creation of dam reservoirs leads to flooding or submerging land and homes, forcing people to relocate. For example, more than 1.4 million people in China were displaced due to the Three Gorges Dam and over 40,000 people in India were displaced due to the Sadar Sarovar Dam. In the twentieth century, an estimated 40–80 million people were displaced by dams worldwide. India is thought to have the greatest number of development-induced displaced people globally. Moreover, the installation of large dams among other developmental projects is the main reason for displacement after the independence of India. Similarly, DVC dams lead to displacement in the DRB. Land, houses, and livelihood were displaced by DVC dams.

5.4.1 Displacement of Local Communities and Tribes

Construction of the DVC dams has led to the substantial displacement of people due to the submergence of land to create reservoirs that impact the communities, especially those living along the river banks with the reservoir's sides in the Damodar River Basin (DRB). For example, the construction of the Panchet and Maithon dams has resulted in the displacement of tribal communities and populations living in the reservoir areas. They combinedly displaced 93,874 people, of whom 56.46% were tribal people (Table 5.12). The construction of these dams displaced large numbers of people from their villages and agricultural lands, causing social and economic upheaval. The indigenous communities were forcibly relocated from their homes and livelihoods due to the acquisition of large tracts of land that were traditionally inhabited. The tribal people, the majority belonging to communities like the Santhal and Munda tribes, have been displaced in the Valley region due to the DVC dam project. The land acquisition from tribal communities without adequate consultation or compensation resulted in the displacement and disruption of their traditional way of life. This displacement leads to serious socio-economic challenges for tribal people, such as loss of access to land, forests, and traditional sources of income, leading to marginalization and vulnerability. Although the dams have been installed for multi-purpose benefits, this displacement is mostly accompanied by the problems of inadequate compensation and rehabilitation. Despite displacement policies, the rehabilitation or resettlement process for tribal communities has mostly been inadequate. Many displaced people were left without proper housing, employment opportunities, and access to basic services.

Table 5.12 Dams and the displacement of tribal people in India

Name of project*	State	Population facing displacement	Tribal people as percentage of displaced
Karjan	Gujarat	11,600	100
Sardar Sarovar	Gujarat	200,000	57.6
Maheshwar	Madhya Pradesh	20,000	60
Bodhghat	Madhya Pradesh	12,700	73.91
Icha	Bihar	30,800	80
Chandil	Bihar	37,600	87.92
Koel Karo	Bihar	66,000	88
Mahi Bajaj Sagar	Rajasthan	38,400	76.28
Polavaram	Andhra Pradesh	150,000	52.9
Maithon & Panchet	Bihar (now Jharkhand)	93,874	56.46
Upper Indravati	Orissa	18,500	89.2
Pong	Himachal Pradesh	80,000	56.25
Inchampalli	Andhra Pradesh–Maharashtra	38,100	76.28
Tultuli	Maharashtra	13,600	51.61
Daman Ganga	Gujarat	8700	48.7
Bhakra	Himachal Pradesh	36,000	34.76
Masan Reservoir	Bihar	3700	31
Ukai Reservoir	Gujarat	52,000	18.92

Source: Hemadri et al. (1999). (*Note: Projects are either under construction or have been planned.)

5.4.2 *Livelihood Crisis and Socio-Economic Challenges of the Displaced*

Displacement significantly affects people's livelihoods and socio-economic well-being, leading to several challenges. These include loss of traditional livelihoods, displacement-induced destitution, poverty, unemployment, and difficulties in meeting basic needs such as shelter, healthcare, education, and other essential services like food, clothing, and clean water. Difficulty in accessing adequate food leads to malnutrition and health problems. It disrupts the housing conditions, health, safety, livelihoods, and education of displaced people. It affects their ability to contribute to the local economy, earn income, pay rent or taxes, purchase goods, and invest in their socio-economic development. Moreover, it disrupts social life, increases the risk of abuse and exploitation, and has long-term consequences for mental health and overall well-being. It forces people to abandon their traditional occupations and access to resources like land and common property, leading to economic hardship, unemployment, and vulnerability.

Forced displacement results in the loss of property, land, other assets, and sources of income or livelihoods, leaving people impoverished and dependent on humanitarian assistance. It also causes the loss of cultural practices, languages, and traditions, affecting cultural identity and heritage. Additionally, displaced populations are often at high risk of infectious diseases due to overcrowding, poor sanitation, and lack of access to healthcare. It leads to unemployment due to restrictions on work permits, a lack of recognized skills, or competition for limited opportunities, making it difficult to find jobs, resulting in individuals often being forced to take informal or low-paying positions. It disrupts social networks, leading to isolation, anxiety, shocks, and stresses, and increasing the risk of violence, which potentially impacts mental health and well-being. The loss of livelihoods and social support makes displaced individuals more vulnerable to abuse, exploitation, and violence. Displacement has long-lasting consequences for the socio-economic development of individuals and communities, potentially locking them into a cycle of poverty and dependency. It disrupts food systems and undermines food sovereignty, affecting access to food and local agricultural activities. The financial costs of internal displacement can be overwhelming for both Internally Displaced Persons (IDPs) and host communities, potentially straining government resources.

DVC dams lead to huge socio-economic development in the basin area, as well as the regional economy by providing water for irrigation, industries and domestic use, and other forms of development such as controlling floods, generation and provision of electricity, creation of a fishing ground, transport and embankment, afforestation, soil erosion check, and recreation area. On the other hand, displacement caused by dams is one of the most significant social consequences, which leads to a negative impact on the livelihoods and socio-economic conditions of displaced people as well as the regional. Despite government efforts, resettlement programmes often face challenges, including inadequate compensation, lack of proper land allocation for displaced communities, and inadequate access to basic amenities.

5.5 Dams, LULC Changes and Related Issues

5.5.1 Erosion and Sediment Load Due to LULC Dynamics

Land use and land cover (LULC) changes specifically affect erosion and sediment loads by shifting surface runoff and soil erosion. Increasing erosion and sediment yield as a result of deforestation, agricultural and urban areas growth and unsustainable practices (Aneseyee et al., 2020; Sadhwani et al., 2022; Nigussie et al., 2025). In addition, mining and reservoir construction cause land cover changes that accelerate sediment transport and river bank erosion, triggering thalweg shifts, channel intersections and gradual changes in river cross-sections (Kayitesi et al., 2022). Although growth in urban areas, cropland and bare soil is generally associated with higher soil loss, forest and grassland cover normally provide better protection by reducing runoff and sediment transport (Aneseyee et al., 2020; Moisa et al., 2021). Vegetation presence and its type play an important role in erosion and sediment load. For example, forests with their extensive root systems and canopy cover decrease the effect of rainfall and encourage water infiltration, resulting in less surface runoff and erosion (Zhao et al., 2017; Kayitesi et al., 2022). However, areas with little vegetation, like degraded land or bare soil, are extremely vulnerable to erosion, leading to a substantial portion of soil loss and sediment transport (Aneseyee et al., 2020; Kayitesi et al., 2022). The expansion of urban areas transforms landforms into impermeable surfaces, which increases surface runoff and sediment export, leading to bank erosion and changing river morphology (Sadhwani et al., 2022; Kayitesi et al., 2022). Soil loss is due to the expansion of agricultural land, especially when combined with practices such as farming and deforestation. However, conservation activities like contour farming and terracing can mitigate this through decreasing soil erosion and facilitating siltation near the source (Aneseyee et al., 2020; Kayitesi et al., 2022). Increased soil erosion in catchment areas increases sediment loads in rivers and reservoirs, which leads to a reduction in their storage capacity and affects their effectiveness for water supply, irrigation, and hydropower (Kayitesi et al., 2022; Bihonegn & Awoke, 2023).

The LULC change in the DRB is mainly due to urbanization, agriculture and mining (Hoque et al., 2022b; 2024; Ghosh et al., 2022), which have notably increased erosion and sediment loads as a result of vegetation removal and soil exposure (Alam & Maiti, 2025a, 2025b). Its changes are directly related to high soil loss rates, and significant areas in the basin are facing very high potential for erosion (Ghosh et al., 2025). Increased erosion in the basin results in large amounts of sediment being exported into river systems (Alam & Maiti, 2025a, 2025b). This change also leads to high contaminant accumulations, disrupted river flow, reduced reservoir capacity, increased flood risk and overall environmental degradation (Goswami et al., 2023). Deforestation for agriculture and the removal of protective vegetation for activities like extensive coal mining in the upper Damodar Basin expose the soil to water and wind, leading to increased erosion (Alam & Maiti, 2025b). Agricultural practices changes may also contribute to increased soil erosion and sediment export (Goswami et al., 2023; Alam & Maiti, 2025a, 2025b). Mining operations and industrial activities in the basin also lead to the release of coal dust, fly ash and heavy metals into the river systems, which degrade water quality along with sedimentation (Hoque et al., 2022a; Hoque & Islam, 2024; Ghosh et al., 2025). Besides, the built-up areas modifies drainage patterns, thereby increasing surface runoff and erosion (Hoque et al., 2025a, b; Goswami et al., 2023). Accumulation of silt in reservoirs decreases their storage capacity and impacts their long-term viability (Ibisate et al., 2025; Alam & Maiti, 2025a, 2025b). A large amount of siltation can block the stream, interrupt natural flow and affect aquatic biota (Goswami et al., 2023; Hoque et al., 2025a, b). Soil erosion with rising sediment export may lead to more frequent and severe flooding in the basin, especially downstream (Alam & Maiti, 2025b). Besides, sediment and associated contaminants degrade water quality, which impacts the ecosystem as well as human health (Islam et al., 2025; Ghosh et al., 2025).

Hence, effective management is required to monitor LULC to develop sustainable catchment areas and sediment control approaches or strategies. Constant monitoring of LULC change is essentially required for understanding, forecasting and predicting sediment dynamics (Alam & Maiti, 2025a, 2025b). Besides, implementing sustainable basin management practices can help control soil erosion and sedimentation. Moreover, integrating techniques such as remote sensing, modelling, and soil conservation systems is critical for effective management (Patro et al., 2022; Ghosh et al., 2025; Alam & Maiti, 2025a, 2025b).

5.5.2 Impact of LULC Changes on Vegetation Changes (Evapotranspiration Status)

LULC change, mainly the transition from vegetated areas such as forest to non-vegetated land or built-up areas, usually reduces evapotranspiration (ET) due to decreased barrier, infiltration, and a smaller available land area for water vapour return to the atmosphere (Abuhay et al., 2023; Debele & Beketie, 2025). This transformation leads to increased surface runoff and flood risk and decreased groundwater recharge (Kayitesi et al., 2022). However, ET also increases significantly if LULC changes occur due to the growth of certain vegetation such as forests and shrublands, which are highly efficient in respiration (Abuhay et al., 2023). Replacing forests with other land, like agriculture or built-up or mining areas, decreases barriers and intrusions (Shukla et al., 2023), which leads to a reduction in ET because there is a smaller amount of vegetation to expose to water vapour and little soil moisture to evaporate (Abuhay et al., 2023). Urban areas have larger impervious surfaces, leading to reductions in ET (Debele & Beketie, 2025). However, they may have some ET from green spaces, but they are less effective than forest areas (Mabuda et al., 2024). Whereas increased forest or shrub cover may result in higher ET rates. For example, such vegetation gives shade, endorses infiltration, and emits vapour into the atmosphere by evaporation, which provides a cooling effect (Abuhay et al., 2023; Debele & Beketie, 2025).

Less water transpiration and evaporation with more precipitation increases the surface runoff, which potentially leads to more frequent and severe flooding (Kayitesi et al., 2022; Abuhay et al., 2023). In addition, reduced infiltration due to land cover change decreases groundwater recharge, which affects groundwater availability (Kayitesi et al., 2022; Abuhay et al., 2023). Thus, large-scale LULC changes, particularly deforestation in the tropical regions, have significant influences on the water cycle by affecting infiltration, flow, and water balance in an ecosystem (Shukla et al., 2023).

LULC changes in the DRB, due to factors such as expansion of urbanization and agricultural land, lead to reduced vegetation cover, which decreases ET and modifies the regional water balance (Das et al., 2018). Moreover, deforestation and changes to built-up or agricultural zones reduce canopy cover, resulting in less transpiration and interception, which leads to a reduction in overall ET and increasing runoff (Biswas, 2014). Several studies indicate that forest and vegetation cover have been reduced over the years in the DRB as a result of built-up and agricultural land expansion (Mishra & Paliwal, 2024). Less forest cover, especially vegetation, reduces interception and evaporation, which directly decreases total evaporation of an area (Das et al., 2018). Less water retention by vegetation and return to the atmosphere due to LULC changes increased runoff and baseflow, leading to a reduction in ET (Das et al., 2018). Transformation of forest land to built-up or agricultural land also reduces soil infiltration, leading to further surface waterlogging, amplified erosion and probable flooding (Biswas, 2014). Thus, changes in vegetation cover have substantial effects on basin hydrology, contributing to amplified erosion and alterations in river morphology and stream discharge.

Expansion of built-up and agricultural land due to reduced natural vegetation is a main driver of this LULC change (Biswas, 2014; Shukla et al., 2023). A massive reduction in forest cover due to deforestation directly contributes to the overall decrease in vegetation cover and associated ET (Das et al., 2018). Heavy mining and associated industrial activities have also caused significant changes in land use and vegetation cover in some parts of the basin (Anshumali et al., 2025). Decreased ET and increased flows impact the overall water balance, leading to significant changes in river discharge patterns (Das et al., 2018). Erosion activity also increases along the riverbank and changes the morphology of the river (Biswas, 2014). Thus, these LULC changes affect the ET regionally, overall hydrologic and also drought prevalence in the basin area (Das et al., 2018; Kumar et al., 2025).

5.5.3 Impact of LULC Changes on Greenhouse Gas (GHG) Production

LULC changes also notably affect greenhouse gas (GHG) production through releasing carbon stored from forests and soils and rising releases from agriculture and other human activities (Berhanu et al., 2023). Increased built-up areas lead to the growth of non-impervious surfaces such as residential areas, industrial areas and road networks. It increases the greenhouse gas emissions like carbon dioxide and methane (Kayitesi et al., 2022; Debele & Beketie, 2025). Deforestation, forests to agriculture, and agriculture-related land degradation are significant sources of GHG emissions; however, reforestation, afforestation, and agricultural practices changes may lead to carbon sequestration, reducing GHG accumulations (Huang et al., 2023). Hence, sustainable agricultural practices may recover carbon sequestration in soil and reduce GHG fluxes (Huang et al., 2023). As LULC changes signify a substantial percentage of global anthropogenic GHG emissions, considering these dynamic forces is critical to managing GHG production (Kearney & Smukler, 2016). Deforestation, especially in the tropical areas, leads to significant loss of soil biomass and carbon, leading to increased CO_2 releases (van Beijma et al., 2018; Berhanu et al., 2023). Besides, the agricultural expansion and intensive practices lead to rising GHG releases by shifting soil carbon stocks (Kearney & Smukler,

2016; Huang et al., 2023). Moreover, land degradation for numerous activities, like the extension of large-scale commodity yields in peatlands, leads to substantial changes in carbon stock (van Beijma et al., 2018; Ateşoğlu et al., 2025). Changes in forests like settlements or urban areas also contribute to rising GHG releases (Larbi, 2023; Ateşoğlu et al., 2025). However, afforestation and reforestation or planting trees may increase surface carbon sequestration and stock, as a result, the concentration of atmospheric CO_2 decreases (Berhanu et al., 2023).

LULC changes in the DRB due to the expansion of built-up and agricultural areas and mining activities, leading to the destruction of forests and grasslands, which significantly contribute to augmented GHG emissions (Kumar et al., 2023). This destruction or loss of forests and grasslands is an important driver of higher carbon releases. With this, the overall net rise in GHG is often linked with the change of vegetation to built-up and agricultural areas. These alterations also impact water quality and flow. Agricultural expansion, especially the use of fertilizers, often leads to high emissions of nitrous oxide and methane, powerful greenhouse gases (Goswami et al., 2023). Expanding built-up areas, particularly increasing energy consumption and industrial activity, contribute to higher GHG production (Punja et al., 2024). Several studies show notable growth in agricultural and built-up areas occurs in the DRB at the cost of vegetation, wastelands and even water bodies (Anshumali et al., 2025; Kumar et al., 2023). Deforestation directly affects the basin's ability to sequester carbon, leading to a net rise in GHG production (Gong et al., 2022; Kumar et al., 2023). Overall, LULC changes performance as a net source of GHG releases for the basin, amplifying the effects of climate change (Khachoo et al., 2024; Ateşoğlu et al., 2025). Increasing GHG emissions from LULC changes contribute to the larger global warming problem. Therefore, addressing these LULC changes with better planning is vital for developing comprehensive strategies for managing carbon stocks and mitigating climate change effects in the basin (Khachoo et al., 2024; Ateşoğlu et al., 2025; Zewdu et al., 2025).

The Land Use, Land Use Change and Forestry (LULUCF) sector has the greatest potential to reduce anthropogenic GHG emissions by reducing deforestation and forest degradation. Several strategies focused on carbon sequestration through forestry, agriculture, and ecosystem restoration are crucial for mitigating climate change (Huang et al., 2023). Remote sensing (RS) and Geographic Information System (GIS), including high-resolution satellite imagery and remote sensing technologies, can be used for monitoring LULC changes and measuring effects linked with GHG emissions, removals, and mitigation (Kearney & Smukler, 2016; Khachoo et al., 2024).

5.5.4 *Impact of the Reservoir on GHG Production*

Reservoirs can produce GHG by decomposing submerged organic matter, producing CO_2 and CH_4, which in turn become more powerful (Lu et al., 2020; Rodríguez & Peñuela, 2022). Nitrous oxide (N_2O) may also be produced, although in lesser amounts. Although CO_2 emissions reduce over time, methane emissions can become more prominent, particularly in warmer climates, and are produced in anaerobic, deep aquifers. Reservoir age, climate, the volume of organic matter, and nutrients in the system are very significant for GHS emissions. Reservoirs in the DRB, built by the DVC, increase GHG production through submerging organic-rich soil, which decomposes to release CO_2 and CH_4. High concentrations of organic matter and nutrients can lead to rising GHG emissions, and decomposition processes are significant for CO_2 and CH_4 production. These gases are released through different ways, such as water surface expansion, ebullition (bumping), and emissions from turbines. The transformation of a natural ecosystem into a reservoir alters biogeochemical processes, leading to higher GHG emissions than a free-flowing river. However, specific impacts depend on reservoir age, depth, and surrounding watershed characteristics, with some studies indicating low to moderate GHG risk for many Indian hydropower reservoirs (Kumar et al., 2021). It is essential to consider GHG emissions during the planning and design of new dam infrastructure. Converting large areas of terrestrial ecosystems to reservoirs can make a significant contribution to their regional and global carbon budget. For sustainable development, it is important to understand these impacts and develop effective policies and mitigation measures for future dam projects.

5.6 Concluding Notes

The creation of the DVC leads to the socio-economic development of the inhabitants of the study area. The developing activities, such as controlling floods, supplying water for irrigation, and industrial and municipal usage in the basin area, by the optimum use of available water resources in the basin. It has developed the agricultural sector by providing water for irrigation and improving irrigation facilities in the study area. The crop diversity, intensity, and productivity results also indicate that the agricultural sector has developed after the DVC project. It controls floods, and flood probability decreases substantially as several devastating floods occurred in the basin area before the DVC dams. It has developed the industrial sector by supplying water for industrial use. It also supplies water for domestic usage. The other activities include strengthening eco-conservation activities, making greenery, constructing roads and embankments, checking soil erosion in the valley area, electricity generation and provision, developing tourism and fish farming in and around the DVC reservoir area, and improving the financial health of the corporation through adopting efficient industrial, commercial and human resource management practices.

Besides socio-economic development, dams cause the displacement of local and tribal communities, which is the most significant social impact of dam construction. They face a livelihood crisis and socio-economic challenges. Moreover, LULC dynamics accelerate erosion and increase sediment loads. In the DRB, changes are mainly driven by urbanisation, agriculture, mining, and industrialisation, which substantially raise soil erosion, riverbank erosion, and sediment loads in the river system. LULC changes, particularly vegetation decrease, affect the evapotranspiration levels in the basin. These changes also influence greenhouse gas (GHG) emissions. Additionally, reservoirs can also emit GHGs within the basin area. Hence, it is necessary to measure dams and environmental sustainability in the basin area, which is taken up in the next chapter, i.e., Chap. 6.

References

Abuhay, W., Gashaw, T., & Tsegaye, L. (2023). Assessing impacts of land use/land cover changes on the hydrology of Upper Gilgel Abbay watershed using the SWAT model. *Journal of Agriculture and Food Research, 12*, 100535. https://doi.org/10.1016/j.jafr.2023.100535

Alam, S. A., & Maiti, R. (2025a). Land use land cover dynamics and sediment variation. In *Sediment source to sink: Deciphering sediment connectivity to large dams in Damodar River Basin* (pp. 129–151). Springer. https://doi.org/10.1007/978-3-031-87072-9_7

Alam, S. A., & Maiti, R. (2025b). Deciphering sediment pathways: Novel integrated approaches for sediment source identification and vulnerability prediction by machine learning models in major dam catchments in Chota Nagpur Plateau, India. *Physics and Chemistry of the Earth, Parts A/B/C*, 103974. https://doi.org/10.1016/j.pce.2025.103974

Altinbilek, D. (2002). The role of dams in development. *International Journal of Water Resources Development, 18*(1), 9–24.

Aneseyee, A. B., Elias, E., Soromessa, T., & Feyisa, G. L. (2020). Land use/land cover change effect on soil erosion and sediment delivery in the Winike watershed, Omo Gibe Basin, Ethiopia. *Science of the Total Environment, 728*, 138776. https://doi.org/10.1016/j.scitotenv.2020.138776

Anshumali, Pandey, R. K., Kumar, G., Prasad, N., Kumar, A., Gogoi, S., et al. (2025). Impact of LULC changes on small watershed morphometry and delineation of sustainable conservation strategies for degraded transboundary rivers. *Scientific Reports, 15*(1), 24063. https://doi.org/10.1038/s41598-025-08841-2

Ateşoğlu, A., Serengil, Y., Şensoy, H., & Bulut, F. Ş. (2025). LULC changes in Türkiye's mountains for the 2000–2022 period and its implications on national GHG emissions. *Environmental Monitoring and Assessment, 197*(7), 715. https://doi.org/10.1007/s10661-025-14136-w

Berhanu, Y., Dalle, G., Sintayehu, D. W., Kelboro, G., & Nigussie, A. (2023). Land use/land cover change-induced greenhouse gas emissions in the global biosphere reserve of tropical Afromontane forest. *Scientific African, 21*, e01811. https://doi.org/10.1016/j.sciaf.2023.e01811

Bhattacharyya, K. (2011). *The Lower Damodar River, India: Understanding the human role in changing fluvial environment*. Springer.

Bhattacharyya, K., & Wiley, M. J. (2014). Dams, riparian settlement and the threat of climate change in a dynamic fluvial environment: A case study of the Damodar River, India. In *Large dams in Asia: Contested environments between technological hydroscapes and social resistance* (pp. 75–100).

Bihonegn, B. G., & Awoke, A. G. (2023). Evaluating the impact of land use and land cover changes on sediment yield dynamics in the upper Awash basin, Ethiopia the case of Koka reservoir. *Heliyon, 9*(12). https://doi.org/10.1016/j.heliyon.2023.e23049

Biswas, S. S. (2014). Influences of changes in land use/land cover and precipitation variability on hydrology and morphology of middle course of Damodar River in East India. *The International Journal of Earth Sciences and Engineering, 7*(1), 326–338.

Damodar Valley Corporation. (2021). *DVC Reservoir Overview*. www.dvc.gov.in/dvcwebsite_newl/damsbarrages/DamodarValleyCoporation

Das, P., Behera, M. D., Patidar, N., Sahoo, B., Tripathi, P., Behera, P. R., et al. (2018). Impact of LULC change on the runoff, base flow and evapotranspiration dynamics in eastern Indian river basins during 1985–2005 using variable infiltration capacity approach. *Journal of Earth System Science, 127*(2), 19. https://doi.org/10.1007/s12040-018-0921-8

Debele, G. B., & Beketie, K. T. (2025). Monitoring the dynamics of land use and land cover, and their impact on seasonal land surface temperature in the Upper Awash Basin, Central Ethiopia. *Discover Applied Sciences, 7*(4), 321. https://doi.org/10.1007/s42452-025-06742-7

DVC. (2013). *Annual Report 2012–13*. Retrieved August 10, 2021, from https://www.dvc.gov.in/dvcwebsite_new1/downloads/annual-report/

DVC. (2016). *Annual Report 2015–16*. Retrieved August 9, 2021, from https://www.dvc.gov.in/dvcwebsite_new1/downloads/annual-report/

DVC. (2020). *Reservoirs and barrages: An overview of water management*. http://www.dvc.gov.in

DVC. (2022). *Annual Report 2021–22*. Retrieved December 18, 2023, from https://www.dvc.gov.in/dvcwebsite_new1/downloads/annual-report/

Ghosh, S., & Guchhait, S. K. (2014). Hydrogeomorphic variability due to dam constructions and emerging problems: A case study of Damodar River, West Bengal, India. *Environment, Development and Sustainability, 16*, 769–796. https://doi.org/10.1007/s10668-013-9494-5

Ghosh, S., & Mistri, B. (2013). Performance of DVC in flood moderation of lower Damodar River India and emergent risk of flood. *Eastern Geographer, 19*(1), 55–66.

Ghosh, P. K., Bandyopadhyay, S., Jana, N. C., & Mukhopadhyay, R. (2016). Sand quarrying activities in an alluvial reach of Damodar River, Eastern India: Towards a geomorphic assessment. *International Journal of River Basin Management, 14*(4), 477–489. https://doi.org/10.1080/15715124.2016.1209509

Ghosh, J., Das, S., Ghosh, S., Midya, K., Das, S., & Mishra, V. N. (2025). Soil erosion vulnerability analysis of Damodar River Basin, India using Revised Universal Soil Loss Equation (RUSLE) in Google Earth Engine (GEE). *Discover Geoscience, 3*(1), 1–31. https://doi.org/10.1007/s44288-025-00191-w

Ghosh, S., Hoque, M. M., Saha, U. D., & Islam, A. (2022). Assessment of dam-induced changes in ecogeomorphological behaviour and fluvial functionality in the Damodar River, West Bengal, India. *AQUA—Water Infrastructure, Ecosystems and Society, 71*(6), 722–750. *https://doi.org/10.2166/aqua.2022.003*

Gong, W., Duan, X., Mao, M., Hu, J., Sun, Y., Wu, G., ... & Liu, T. (2022). Assessing the impact of land use and changes in land cover related to carbon storage by linking trajectory analysis and InVEST models in the Nandu River Basin on Hainan Island in China. *Frontiers in Environmental Science, 10*, 1038752. https://doi.org/10.3389/fenvs.2022.1038752

Goswami, G., Mandal, S., Basack, S., Mukherjee, R., & Karakouzian, M. (2023). Assessing the impacts of land use and land cover changes on the water quality of River Hooghly, West Bengal, India: A case study. *Hydrology, 10*(3), 71. https://doi.org/10.3390/hydrology10030071

Hoque, M. M., Islam, A., & Ghosh, S. (2022a). Environmental flow in the context of dams and development with special reference to the Damodar Valley Project, India: a review. *Sustainable Water Resources Management, 8*(3), 1–27. https://doi.org/10.1007/s40899-022-00646-9

Hoque, M. M., Islam, A., & Mahammad, S. (2022b). Assessing the surface and bottom river water quality for drinking purpose and human health risk analysis: a study of Damodar River, India. *Arabian Journal of Geosciences, 15*(23), 1734. https://doi.org/10.1007/s12517-022-10987-6

Hoque, M. M., Islam, A., Islam, A. R. M. T., Pal, S. C., Mahammad, S., & Alam, E. (2023). Assessment of soil heavy metal pollution and associated ecological risk of agriculture dominated mid-channel bars in a subtropical river basin. *Scientific Reports, 13*(1), 11104. https://doi.org/10.1038/s41598-023-38058-0

Hoque, M. M., & Islam, A. (2024). Spatio-temporal dynamics of ecological, bacteriological, and overall water quality of the Damodar River, India. *Environmental Science and Pollution Research, 31*(12), 18465–18484. https://doi.org/10.1007/s11356-024-32185-5

Hoque, M. M., Islam, A., Sarkar, B., & Saha, U. D. (2024). Assessing the surface and bottom river water quality for irrigation: A study of Damodar River, India. *International Journal of Energy and Water Resources, 8*(4), 545–562. https://doi.org/10.1007/s42108-022-00206-z

Hoque, M. M., Islam, A., Pal, S. C., Sarkar, B., & Hossain, Z. M. (2025a). Assessment of fish habitat suitability in the wake of elevated pollution of a tropical river in India. *Environmental Science and Pollution Research, 32*(3), 1208–1222. https://doi.org/10.1007/s11356-024-35685-6

Hoque, M. M., Pal, S. C., Mandal, S., & Islam, A. (2025b). Assessment of mollusc habitat suitability of an anthropogenically altered tropical river in India. *Environmental Monitoring and Assessment, 197*(10), 1–18. https://doi.org/10.1007/s10661-025-14593-3

Hemadri, R., Mander, H., & Nagaraj, V. (1999). *Dams, displacement, policy and law in India*. World Commission on Dams.

Huang, S., Ghazali, S., Azadi, H., Moghaddam, S. M., Viira, A. H., Janečková, K., et al. (2023). Contribution of agricultural land conversion to global GHG emissions: A meta-analysis. *Science of The Total Environment, 876*, 162269. https://doi.org/10.1016/j.scitotenv.2023.162269

Ibisate, A., García, H., Vázquez-Tarrío, D., Sánchez-Pinto, I., Herrero, X., de Olazagoitia, A. S., & Ollero, A. (2025). Bedload response to dam removal: Results from a 6-year particle tracking survey in the Leitzaran River (Basque Country). *Geomorphology, 470*, 109542. https://doi.org/10.1016/j.geomorph.2024.109542

Islam, A., Sardar, N., Mohinuddin, S., Hoque, M. M., Sengupta, S., Das, B. C., ... & Sengupta, B. (2023). Quasi-equilibrium channel metamorphosis in planform of a subtropical river in India in post-dam period. *Catena, 221*, 106793. https://doi.org/10.1016/j.catena.2022.106793

Islam, A., Das, B. C., Mahammad, S., Hoque, M. M., Pham, Q. B., Sarkar, B., ... & Barman, S. D. (2025). Assessing river water quality for ecological risk in the context of a decaying river in India. *Environmental Science and Pollution Research, 32*(16), 10223–10239. https://doi.org/10.1007/s11356-024-33684-1

Kayitesi, N. M., Guzha, A. C., & Mariethoz, G. (2022). Impacts of land use land cover change and climate change on river hydro-morphology—A review of research studies in tropical regions. *Journal of Hydrology, 615*, 128702. https://doi.org/10.1016/j.jhydrol.2022.128702

Kearney, S. P., & Smukler, S. M. (2016). Determining greenhouse gas emissions and removals associated with land-use and land-cover change. In *Methods for measuring greenhouse gas balances and evaluating mitigation options in smallholder agriculture* (pp. 37–70).

Khachoo, Y. H., Cutugno, M., Robustelli, U., & Pugliano, G. (2024). Impact of land use and land cover (LULC) changes on carbon stocks and economic implications in Calabria using Google Earth Engine (GEE). *Sensors, 24*(17), 5836. https://doi.org/10.3390/s24175836

Khorakiwala, A. A. (2017). *The well-fed subject: Modern architecture in the quantitative state, India (1943-1984)* (Doctoral dissertation, Harvard University).

Kumar, A., Yu, Z. G., Klemeš, J. J., & Bokhari, A. (2021). A state-of-the-art review of greenhouse gas emissions from Indian hydropower reservoirs. *Journal of Cleaner Production, 320*, 128806. https://doi.org/10.1016/j.jclepro.2021.128806

Kumar, S., Mishra, A., & Singh, U. K. (2023). Assessment of land cover changes and climate variability effects on catchment hydrology using a physically distributed model. *Sustainability, 15*(13), 10304. https://doi.org/10.3390/su151310304

Kumar, A., Kumar, A., Baghel, S., Prasad, V., & Singh, M. C. (2025). Impact of LULC changes on drought propagation in Mand Catchment, Chhattisgarh. *Journal of Applied Water Engineering and Research*, 1–16. https://doi.org/10.1080/23249676.2025.2507741

Larbi, I. (2023). Land use-land cover change in the Tano basin, Ghana and the implications on sustainable development goals. *Heliyon, 9*(4). https://doi.org/10.1016/j.heliyon.2023.e14859

Loomis, J. (2002). Quantifying recreation use values from removing dams and restoring free-flowing rivers: A contingent behavior travel cost demand model for the Lower Snake River. *Water Resources Research, 38*(6), 2–1.

Lu, S., Dai, W., Tang, Y., & Guo, M. (2020). A review of the impact of hydropower reservoirs on global climate change. *Science of the Total Environment, 711*, 134996. https://doi.org/10.1016/j.scitotenv.2019.134996

Mabuda, M. O., Shoko, C., Dube, T., & Mazvimavi, D. (2024). An analysis of the effects of changes in land use and land cover on runoff in the Luvuvhu catchment, South Africa. *Remote Sensing Applications: Society and Environment, 33*, 101144. https://doi.org/10.1016/j.rsase.2024.101144

Majumder, M., Roy, P., & Mazumdar, A. (2010). An introduction and current trends of Damodar and Rupnarayan River Network. In *Impact of climate change on natural resource management* (pp. 461–480). https://doi.org/10.1007/978-90-481-3581-3_25

Mishra, R., & Paliwal, I. K. (2024). Detection of LULC change with erosion and accretion in the catchment area of Panchet Reservoir, India, in the context of sustainable development and management. In *Sustainable local development for environmental and social sustainability* (pp. 141–154). Springer. https://doi.org/10.1007/978-3-031-67303-0_9

Moisa, M. B., Negash, D. A., Merga, B. B., & Gemeda, D. O. (2021). Impact of land-use and land-cover change on soil erosion using the RUSLE model and the geographic information system: A case of Temeji watershed, Western Ethiopia. *Journal of Water and Climate Change, 12*(7), 3404–3420. https://doi.org/10.2166/wcc.2021.131

Nigussie, W., Al-Najjar, H., Kalantar, B., Nega, W., Bizelk, T., Suryabhagavan, K. V., & Adane, N. (2025). Evaluating the impact of land use/land cover and rainfall changes on regional soil erosion. *Geomatics, Natural Hazards and Risk, 16*(1), 2520374. https://doi.org/10.1080/19475705.2025.2520374

Overton, I. C., Smith, D. M., Dalton, J., Barchiesi, S., Acreman, M. C., Stromberg, J. C., & Kirby, J. M. (2014). Implementing environmental flows in integrated water resources management and the ecosystem approach. *Hydrological Sciences Journal, 59*(3–4), 860–877.

Patro, E. R., De Michele, C., Granata, G., & Biagini, C. (2022). Assessment of current reservoir sedimentation rate and storage capacity loss: An Italian overview. *Journal of Environmental Management, 320*, 115826. https://doi.org/10.1016/j.jenvman.2022.115826

Punja, V., Pathak, V., Shukla, S. P., & Tripathi, R. P. (2024). The impact of land use/land cover change on water quality in the middle ganga region: A review. *Topics in Catalysis, 67*(15), 1024–1037. https://doi.org/10.1007/s11244-024-01996-8

Richter, B. D., & Thomas, G. A. (2007a). Restoring environmental flows by modifying dam operations. *Ecology and Society, 12*(1), 1–26.

Richter, B., & Thomas, G. (2007b). *Dam good operations*. https://www.waterpowermagazine.com/features/featuredam-good-operations

Rodríguez, D. C., & Peñuela, G. A. (2022). Estimation of greenhouse gas emissions of a tropical reservoir in Colombia. *Journal of Water and Climate Change, 13*(2), 872–888. https://doi.org/10.2166/wcc.2022.330

Sadhwani, K., Eldho, T. I., Jha, M. K., & Karmakar, S. (2022). Effects of dynamic land use/land cover change on flow and sediment yield in a monsoon-dominated tropical watershed. *Water, 14*(22), 3666. https://doi.org/10.3390/w14223666

Shukla, S., Meshesha, T. W., Sen, I. S., Bol, R., Bogena, H., & Wang, J. (2023). Assessing impacts of Land Use and Land Cover (LULC) change on stream flow and runoff in Rur Basin, Germany. *Sustainability, 15*(12), 9811. https://doi.org/10.3390/su15129811

Tortajada, C. (2014). Dams: An essential component of development. *Journal of Hydrologic Engineering, 20*(1), A4014005-1–A4014005-9.

van Beijma, S., Chatterton, J., Page, S., Rawlings, C., Tiffin, R., & King, H. (2018). The challenges of using satellite data sets to assess historical land use change and associated greenhouse gas emissions: A case study of three Indonesian provinces. *Carbon Management, 9*(4), 399–413. https://doi.org/10.1080/17583004.2018.1511383

WCD. (2000). *Dams and development: A new framework for decisionmaking*. Earthscan Publications Ltd.

Zewdu, D., Krishnan, C. M., Raj, P. N., Makadi, Y. C., Barati, M. K., & Arlikatti, S. (2025). Interactions of land use, land cover, and climate change: A case study of Raichur District, Karnataka, India. *Environmental Challenges*, 101166. https://doi.org/10.1016/j.envc.2025.101166

Zhao, G., Kondolf, G. M., Mu, X., Han, M., He, Z., Rubin, Z., et al. (2017). Sediment yield reduction associated with land use changes and check dams in a catchment of the Loess Plateau, China. *Catena, 148*, 126–137. https://doi.org/10.1016/j.catena.2016.05.010

Chapter 6
Dams and Environmental Sustainability

6.1 Backdrop: Global Overview

Dam construction is one of the most debatable issues in the world as it promotes development on one side and degrades the aquatic ecosystems on the other (Hoque et al., 2022a). Primarily, it is installed for sustainable water resource management for socio-economic benefits and environmental sustainability (WCD, 2000). It prompts socio-economic development by supplying water and hydropower for domestic, irrigation, and industrial uses (Altinbilek, 2002; Loomis, 2002). Moreover, it provides benefits from hydropower generation, water usage, irrigation, flood control, drought mitigation, flow regulation, fisheries, recreation, and navigation (WCD, 2000; Ghosh & Mistri, 2013; Tortajada, 2015). In this context, DVC dams have been installed in the basin area and DVC manifests massive socio-economic benefits by controlling the colossal floods, developing irrigation systems, supplying domestic and industrial water, generating hydropower, checking soil erosion, boosting fishing, recreation tourism and also increasing vegetation cover (Damodar Valley Corporation, 2021). However, the dams and barrages construction has significantly altered the environmental flow in terms of flow duration, flood frequency, and magnitude of the river systems (Hoque et al., 2025b). The rapid urban-industrial growth in the basin area has increased the pollution of the river water (Hoque et al., 2022b). This typical alteration in the flow characteristics and water quality has altered the eco-geomorphological behaviour of the river systems (Ghosh et al., 2022).

M. M. Hoque, A. Islam, *Dams, Development and Environmental Flow in the Damodar River Basin*, https://doi.org/10.1007/978-3-032-11161-6_6

6.2 Dams and Environmental Sustainability in the Damodar River Basin

The Damodar Valley Corporation (DVC) or Damodar Valley Project (DVP) is a multi-purpose valley project in India, established in 1948 for regional socio-economic development. After constructing four dams (Tilaiya, Maithan, Panchet, and Konar) and one barrage (Durgapur), the DVC authority has a great manifesto regarding socio-economic development in the valley as controlling flood, providing water, although, there are so many negative impacts on the environment. The DVC, economic development, and environmental sustainability are integrated in a bi-directional manner, i.e., the DVC promotes economic development on the one hand and interrupts ecological functionality on the other. Thus, it is necessary to assess the impact of DVP and its sustainability. The rapid impact assessment matrix (RIAM) is such a tool that can detect the positive and negative dimensions of the fluvial systems in the wake of dam construction. RIAM is a technique used to measure a project's short- or long-term impact assessment. This chapter justifies the implications of the DVC and local perceptions regarding environmental sustainability. This is achieved through the utilization of the RIAM method and a mathematical sustainability model. The database, methods, and analytical design are discussed in the following sections.

6.2.1 RIAM Model

A total of 68 parameters have been selected for analysing DVC, development, and environmental sustainability. Out of 68 parameters, 30 for physical/chemical, 17 for biological/ecological, 14 for social/cultural and 7 for economic/operational have been considered for the rapid impact assessment matrix (RIAM). The components selection for RIAM is given in Table 6.1. A total of 7 upstream and 5 downstream villages from Panchet and Maithon dam (as a terminal dam on the Damodar River) have been selected for the RIAM analysis. The village selection details are discussed in the earlier chapter (Chap. 5).

The rapid impact assessment matrix (RIAM) technique is used to assess the sustainability of a particular project. It is a semi-quantitative method used to determine the assessment criteria and deliver an accurate and independent score for each state. In this method, the quantitative and qualitative data are assembled to assign an independent score for each condition. The scores of each aspect are calculated by five assessment criteria, such as the importance of impact (A1), magnitude (A2), permanence (B1), reversibility (B2), and cumulatively (B3), based on Pastakia and Jensen (1998). In this method, the environmental components are categorized into four: (1) physical/chemical, (2) biological/ecological, (3) social/cultural, and (4) economic/operational. The criteria are categorized into two: (A) significant to the condition that affects the final score individually, and (B) demonstrates the significance of the

Table 6.1 Component details for the rapid impact assessment matrix

ID	Category	Factor	Comments
1. Physical/chemical components			
PC 1	Air	Odour	Human health
PC 2		Noise	
PC 3		Vibration	
PC 4	Water	Heavy metals:	Molecular bonding in soils, human health, flora, fauna, aquatic life
PC 5		Mercury (Hg)	
PC 6		Arsenic (As)	
PC 7		Lead (Pb)	
PC 8		Iron (Fe)	
PC 9		Copper (Cu)	
PC 10		Faecal coliforms	Water quality, aquatic life, human health
PC 11		Nutrients	Water quality, aquatic life
PC 12		(BOD)	
PC 13		(COD)	
PC 14		(TOC)	
PC 15		DO	
PC 16		Inundation of lands	Landscape, flora, fauna, soil, socio-economic
PC 17		Basin hydraulic loss	Aquatic life, landscape, soil
PC 18		Streamflow variation	Aquatic life, landscape, erosion/sedimentation, flora, fauna
PC 19		Changes to estuaries	Aquatic life, flora, fauna, landscape, erosion
PC 20		Sedimentation	Aquatic life, water quality
PC 21		Scouring	Aquatic life
PC 22		Turbidity	Aquatic life, water quality
PC 23		Change in pH	
PC 24	Climate	Changes in surface water temperature	
PC 25		Changes in humidity	
PC 26	Soil	Bank erosion	Changes in the water system
PC 27	Landscape	Land use changes	
PC 28		Visual aspects	
PC 29		Physical composition	
PC 30		Impact on sensitive lands	

(continued)

Table 6.1 (continued)

ID	Category	Factor	Comments
2. Biological/ecological components			
BE 1	Flora	Changes in natural vegetation	Reservoir, changes in water system
BE 2		Disturbance of aquatic habitat	
BE 3		Disturbance of plant habitat	
BE 4		Disturbance of natural vegetation	
BE 5		Decrease in biodiversity	
BE 6		Impact of threatened species	
BE 7		Changes in species population	
BE 8		Changes in the aquatic food web	
BE 9		Changes in the mammal food web	
BE 10		Impact on protected areas	
BE 11	Fauna	Migratory changes—fish	
BE 12		Migratory changes—mammals	
BE 13		Disturbance of wildlife habitat	
BE 14		Decrease in biodiversity	
BE 15		Impact on threatened species	
BE 16		Changes in species population	
BE 17		Changes in the mammal food web	

Table 6.1 (continued)

ID	Category	Factor	Comments
3. Sociological/cultural components			
SC 1	Historical monuments	Historical sites	Changes in water levels
SC 2		Archaeological changes	
SC 3		Palaeontological changes	
SC 4	Cultural heritage	Cultural changes	
SC 5		Land use changes	
SC 6		Way of life	
SC 7	Human health and safety	Changes in disease incidence	
SC 8		Increase in parasitic disease	Increase humidity along the shoreline, changes in natural vegetation
SC 9		Risk of surface water contamination	Heavy metals in intermolecular bonding of soils, other pollutants
SC 10	Quality of life	Changes to the well-being of life	Maybe (US) or (DS) villages
SC 11		Changes to quality of life	
SC 12		Quality of recreational facilities	
SC 13		Quantity of recreational facilities	
SC14		Spiritual	
4. Economical/operational components			
EO 1	Natural resources	Present use of natural resources	Maybe (US) or (DS) villages
EO 2		Potential use of natural resources	Increase recreation, fishing, swimming
EO 3	Employment	Employment opportunity	Recreation facilities, dam site
EO 4	Economic development	Economic development-transboundary	Hydro-electricity sales, recreational opportunities
EO 5			Irrigation/ agricultural
EO 6			Industrial development
EO 7			Flood control

situation that does not affect the final score individually (Table 6.2). The score of each aspect is computed using Eqs. 6.1–6.3.

$$(\mathrm{A1})\times(\mathrm{A2})=\mathrm{AT} \tag{6.1}$$

$$(\mathrm{B1})+(\mathrm{B2})+(\mathrm{B3})=\mathrm{BT} \tag{6.2}$$

$$(\mathrm{AT})\times(\mathrm{BT})=\mathrm{ES} \tag{6.3}$$

where the score of AT is calculated based on (A1) and (A2) scores mentioned in criteria A, and the score of BT is calculated based on (B1) and (B2), and (B3) scores

Table 6.2 Description of assessment criteria

Scale	Importance of condition (A1)		Magnitude of change/effect (A2)
Group (A) criteria			
−3	–		Major dis-benefit or change
−2	–		Significant negative dis-benefit or change
−1	–		Negative change to the status quo
0	No importance		No change/status quo
+1	Important only to the local condition		Improvement in status quo
+2	Important to areas immediately outside the local condition		Significant improvement in status quo
+3	Important to regional/national interests		Major positive benefit
+4	Important to national/international interests		–
Group (B) criteria scale	Description		
	Permanence (B1)	Reversibility (B2)	Cumulative (B3)
1	No change/not applicable		
2	Temporary	Reversible	Non-cumulative/single
3	Permanent	Irreversible	Cumulative/synergistic

Based on Batisha (2015)

mentioned in criteria B. The environmental score (ES) is calculated using Eq. 6.3 (Zand & Heir, 2022; Pastakia & Jensen, 1998).

6.2.2 Mathematical Model of Sustainability

This mathematical model was developed by Phillips (2009) for determining the quantitative sustainability from environmental impact assessment (EIA) based data. This model application, using the collected original data in the EIA method, suggests whether the present or proposed project is sustainable or unsustainable. The project's sustainability is considered based on the nature and level of sustainability that are computed by this model.

Four components are used in the RIAM method and their calculated score are used for sustainability assessment. In the present study, the impact of the dams on upstream and downstream villages is evaluated by using this sustainability model to recognize which side (upstream/downstream) is more sustainable/unsustainable in terms of dam, development, and environment.

The sustainability of the present project (DVC) is evaluated using the following mathematical steps:

Step 1: Based on the RIAM score, the relative ES and totals are calculated for upstream and downstream, **Step 2:** Compute specified values for the environment (E) and human needs and interests (HNI) for upstream and downstream,

Step 3: Determination of sustainability or unsustainability based on calculated values of E and HNI for upstream and downstream, **Step 4:** If sustainability (S) is supposed to occur, define the specified S value and S level for upstream and downstream.

Philips's sustainability model is used to delineate whether sustainable development occurs or not for the DVC project using Eq. 6.4.

$$S\text{-value} = E - \text{HNI}, \tag{6.4}$$

where *S* denotes the sustainability, *E* denotes the environmental components (physical, chemical, biological, and ecological) and HNI denotes the human needs and resources (social, cultural, economic, and operational). First of all, the impact assessment results of DVC dams on upstream and downstream are obtained by using the RIAM method (Tables 6.3 and 6.4).

The negative impacts of the DVC dams on environmental components, human needs, and resources are represented in negative numbers, but to derive a positive value for sustainability, add the constant value of 108 to all the obtained values or ES, as suggested in this model, because $108 \leq \text{ES} \leq 108$. This shifts the range of ES values to $0 \leq \text{ES} \leq 216$ and creates the RIAM evaluation results applicable to determine sustainability. Table 6.5 provides the relative ES for the project according to the original ES of upstream and downstream villages used in Tables 6.3 and 6.4.

E and HNI are obtained using Eqs. 6.5–6.6, and Eq. 6.4 yields the level and nature of sustainability.

$$E\text{-value} = \frac{\sum \text{PC} + \sum \text{BE}}{\text{PC}_{\text{max}} + \text{BE}_{\text{max}}} \tag{6.5}$$

$$\text{HNI-value} = \frac{(\text{SC}_{\text{max}} - \sum \text{SC}) + (\text{EO}_{\text{max}} - \sum \text{EO})}{(\text{SC}_{\text{max}} + \text{EO}_{\text{max}})} \tag{6.6}$$

The calculated *E* value is greater than the HNI value indicating the project sustainability and otherwise unsustainability as mentioned in equations.

$$E\text{-value} \geq \text{HNI-value} \Leftrightarrow S > 0,$$
$$E\text{-value} \leq \text{HNI-value} \Leftrightarrow S \leq 0.$$

Finally, the project (DVC) sustainability/unsustainability with their level and nature is evaluated based on Table 6.6.

Table 6.3 Environmental score (ES) for upstream villages

ID	A1	A2	AT	B1	B2	B3	BT	ES
1. Physical/chemical components								
PC 1	2	−2	−4	2	2	3	7	−28
PC 2	1	−1	−1	2	2	3	7	−7
PC 3	1	−1	−1	2	2	2	6	−6
PC 4	3	−3	−9	2	2	3	7	−63
PC 5	3	−2	−6	3	3	3	9	−54
PC 6	3	−2	−6	3	3	3	9	−54
PC 7	3	−2	−6	3	3	3	9	−54
PC 8	3	−1	−3	3	3	3	9	−27
PC 9	3	−1	−3	3	3	3	9	−27
PC 10	3	−2	−6	2	2	3	7	−42
PC 11	3	−3	−9	3	2	3	8	−72
PC 12	3	−1	−3	2	2	3	7	−21
PC 13	3	−1	−3	2	2	3	7	−21
PC 14	3	−1	−3	1	2	2	5	−15
PC 15	3	−2	−6	2	2	3	7	−42
PC 16	1	−2	−2	3	3	3	9	−18
PC 17	3	−3	−9	2	2	3	7	−63
PC 18	3	−3	−9	2	2	3	7	−63
PC 19	3	−3	−9	3	3	3	9	−81
PC 20	3	−3	−9	3	2	3	8	−72
PC 21	0	0	0	1	3	2	6	0
PC 22	3	−2	−6	2	2	2	6	−36
PC 23	3	−1	−3	2	2	2	6	−18
PC 24	1	−1	−1	2	2	2	6	−6
PC 25	0	0	0	2	3	2	7	0
PC 26	0	0	0	1	1	1	3	0
PC 27	3	−3	−9	3	2	3	8	−72
PC 28	2	−1	−2	2	2	2	6	−12
PC 29	3	−2	−6	2	2	3	7	−42
PC 30	3	−1	−3	2	2	3	7	−21
2. Biological/ecological components								
BE 1	3	−2	−6	2	2	2	6	−36
BE 2	3	−2	−6	2	2	3	7	−42
BE 3	3	−1	−3	2	2	2	6	−18
BE 4	3	−2	−6	3	2	2	7	−42
BE 5	3	−2	−6	3	2	3	8	−48
BE 6	3	−2	−6	2	2	2	6	−36
BE 7	3	−2	−6	3	2	3	8	−48
BE 8	3	−2	−6	3	3	3	9	−54
BE 9	3	−2	−6	3	3	3	9	−54

Table 6.3 (continued)

ID	A1	A2	AT	B1	B2	B3	BT	ES
BE 10	3	−2	−6	3	2	3	8	−48
BE 11	3	−2	−6	3	3	3	9	−54
BE 12	3	−2	−6	3	3	3	9	−54
BE 13	3	−2	−6	3	3	2	8	−48
BE 14	3	−3	−9	3	3	3	9	−81
BE 15	3	−3	−9	3	3	3	9	−81
BE 16	3	−2	−6	3	2	3	8	−48
BE 17	3	−2	−6	3	3	2	8	−48
3. Sociological/cultural components								
SC 1	3	1	3	3	3	2	8	24
SC 2	3	−1	−3	3	3	3	9	−27
SC 3	2	−1	−2	3	3	2	8	−16
SC 4	3	2	6	3	3	3	9	54
SC 5	3	−2	−6	3	2	3	8	−48
SC 6	3	2	6	3	3	3	9	54
SC 7	1	1	1	1	1	1	3	3
SC 8	1	1	1	2	2	2	6	6
SC 9	3	−3	−9	2	2	3	7	−63
SC 10	3	3	9	3	3	3	9	81
SC 11	3	3	9	3	3	3	9	81
SC 12	3	2	6	3	3	2	8	48
SC 13	3	3	9	3	3	2	8	72
SC14	2	2	4	3	2	2	7	28
4. Economical/operational components								
EO 1	3	3	9	3	3	3	9	81
EO 2	3	3	9	3	3	3	9	81
EO 3	3	3	9	3	3	3	9	81
EO 4	3	3	9	3	3	3	9	81
EO 5	3	3	9	3	3	3	9	81
EO 6	3	3	9	3	3	3	9	81
EO 7	1	1	1	2	2	2	6	6

6.2.3 *RIAM Model Findings*

According to the DVC authority, the DVC dams have a great positive impact on the socio-economic condition of the DRB as they provide many beneficial sides like controlling floods, providing water for industry, irrigation and domestic use, hydro-electricity generation and creation of fishing grounds and recreation areas. However, they have so many negative impacts on the river systems as well as the basin environment due to diversion and alteration of flow and pollution. Hence, the impact of DVP or DVC dams has been assessed through RIAM based on four environmental assessment categories and the environmental condition of the Damodar River Valley. From this study, it is observed that there are many positive impacts of DVC dams on

Table 6.4 Environmental score (ES) for downstream villages

ID	A1	A2	AT	B1	B2	B3	BT	ES
1. Physical/chemical components								
PC 1	2	−2	−4	2	2	3	7	−28
PC 2	2	−2	−4	2	2	3	7	−28
PC 3	1	−1	−1	2	2	2	6	−6
PC 4	3	−3	−9	3	2	3	8	−72
PC 5	3	−2	−6	3	3	3	9	−54
PC 6	3	−2	−6	3	3	3	9	−54
PC 7	3	−2	−6	3	3	3	9	−54
PC 8	3	−1	−3	3	3	3	9	−27
PC 9	3	−1	−3	3	3	3	9	−27
PC 10	3	−3	−9	3	2	3	8	−72
PC 11	3	−3	−9	3	2	3	8	−72
PC 12	3	−3	−9	3	2	3	8	−72
PC 13	3	−3	−9	3	2	3	8	−72
PC 14	3	−2	−6	2	2	2	6	−36
PC 15	3	−3	−9	3	2	3	8	−72
PC 16	3	−2	−6	2	2	2	6	−36
PC 17	3	−3	−9	3	2	3	8	−72
PC 18	3	−3	−9	3	2	3	8	−72
PC 19	3	−3	−9	3	3	3	9	−81
PC 20	3	−3	−9	3	2	3	8	−72
PC 21	1	−1	−1	3	2	2	7	−7
PC 22	3	−3	−9	2	2	2	6	−54
PC 23	3	−2	−6	2	2	2	6	−36
PC 24	2	−2	−4	2	2	2	6	−24
PC 25	0	0	0	1	1	1	3	0
PC 26	3	−2	−6	2	2	2	6	−36
PC 27	3	−3	−9	3	3	3	9	−81
PC 28	2	−1	−2	3	2	3	8	−16
PC 29	3	−3	−9	3	2	3	8	−72
PC 30	2	−1	−2	2	2	2	6	−12
2. Biological/ecological components								
BE 1	3	−3	−9	3	2	3	8	−72
BE 2	3	−3	−9	3	2	3	8	−72
BE 3	3	−2	−6	3	2	3	8	−48
BE 4	3	−2	−6	3	2	3	8	−48
BE 5	3	−3	−9	3	2	3	8	−72
BE 6	3	−3	−9	3	2	3	8	−72
BE 7	3	−3	−9	3	2	3	8	−72
BE 8	3	−3	−9	3	3	3	9	−81
BE 9	3	−2	−6	3	3	3	9	−54

Table 6.4 (continued)

ID	A1	A2	AT	B1	B2	B3	BT	ES
BE 10	3	−2	−6	3	2	3	8	−48
BE 11	3	−3	−9	3	3	3	9	−81
BE 12	3	−3	−9	3	3	3	9	−81
BE 13	3	−2	−6	3	3	2	8	48
BE 14	3	−3	−9	3	3	3	9	−81
BE 15	3	−3	−9	3	3	3	9	−81
BE 16	3	−3	−9	3	2	3	8	−72
BE 17	3	−2	−6	3	3	2	8	−48
3. Sociological/cultural components								
SC 1	1	1	1	2	2	2	6	6
SC 2	3	−1	−3	2	2	2	6	18
SC 3	2	−1	−2	1	1	1	3	−6
SC 4	3	3	9	3	3	3	9	81
SC 5	3	−3	−9	3	3	3	9	−81
SC 6	3	2	6	3	3	3	9	54
SC 7	2	2	4	1	1	1	3	12
SC 8	1	1	1	2	2	2	6	6
SC 9	3	−3	−9	3	2	3	8	−72
SC 10	3	3	9	3	3	3	9	81
SC 11	3	3	9	3	3	3	9	81
SC 12	2	1	2	3	3	2	8	16
SC 13	1	1	1	3	3	2	8	8
SC14	1	1	1	2	2	2	6	6
4. Economical/operational components								
EO 1	3	3	9	3	3	3	9	81
EO 2	3	3	9	3	3	3	9	81
EO 3	3	3	9	3	3	3	9	81
EO 4	3	2	6	3	3	3	9	54
EO 5	3	3	9	3	3	3	9	81
EO 6	3	3	9	3	3	3	9	81
EO 7	3	3	9	2	2	2	6	54

socio-economic changes, while there are many negative impacts on the environment in the valley region. The DVC dams are used as a water resource management and developmental tool, which positively impacts the economic and operational components like optimum use of natural resources, employment generation and economic development. Besides, they also positively impact the quality of life in socio-cultural components. These positive changes lead to socio-economic development and increase the quality of life. However, they harm the physical and chemical components like air, water and soil with climate and landscape. They also negatively impact on flora and fauna of biological and ecological components due to altered flow and increasing pollution. The details of the environmental assessment score are given in Tables 4.3 and 4.4.

Table 6.5 The relatives ES for the project based on the original ES of upstream and downstream

	Upstream		Downstream	
ID	Original ES	Relative ES	Original ES	Relative ES
PC 1	−28	80	−28	80
PC 2	−7	101	−28	80
PC 3	−6	102	−6	102
PC 4	−63	45	−72	36
PC 5	−54	54	−54	54
PC 6	−54	54	−54	54
PC 7	−54	54	−54	54
PC 8	−27	81	−27	81
PC 9	−27	81	−27	81
PC 10	−42	66	−72	36
PC 11	−72	36	−72	36
PC 12	−21	87	−72	36
PC 13	−21	87	−72	36
PC 14	−15	93	−36	72
PC 15	−42	66	−72	36
PC 16	−18	90	−36	72
PC 17	−63	45	−72	36
PC 18	−63	45	−72	36
PC 19	−81	27	−81	27
PC 20	−72	36	−72	36
PC 21	0	108	−7	101
PC 22	−36	72	−54	54
PC 23	−18	90	−36	72
PC 24	−6	102	−24	84
PC 25	0	108	0	108
PC 26	0	108	−36	72
PC 27	−72	36	−81	27
PC 28	−12	96	−16	92
PC 29	−42	66	−72	36
PC 30	−21	87	−12	96
BE 1	−36	72	−72	36
BE 2	−42	66	−72	36
BE 3	−18	90	−48	60
BE 4	−42	66	−48	60
BE 5	−48	60	−72	36
BE 6	−36	72	−72	36
BE 7	−48	60	−72	36
BE 8	−54	54	−81	27
BE 9	−54	54	−54	54
BE 10	−48	60	−48	60

Table 6.5 (continued)

	Upstream		Downstream	
ID	Original ES	Relative ES	Original ES	Relative ES
BE 11	−54	54	−81	27
BE 12	−54	54	−81	27
BE 13	−48	60	−48	60
BE 14	−81	27	−81	27
BE 15	−81	27	−81	27
BE 16	−48	60	−72	36
BE 17	−48	60	−48	60
SC 1	24	132	6	114
SC 2	−27	81	−18	90
SC 3	−16	92	−6	102
SC 4	54	162	81	189
SC 5	−48	60	−81	27
SC 6	54	162	54	162
SC 7	3	111	12	120
SC 8	6	114	6	114
SC 9	−63	45	−72	36
SC 10	81	189	81	189
SC 11	81	189	81	189
SC 12	48	156	16	124
SC 13	72	180	8	116
SC14	28	136	6	114
EO 1	81	189	81	189
EO 2	81	189	81	189
EO 3	81	189	81	189
EO 4	81	189	54	162
EO 5	81	189	81	189
EO 6	81	189	81	189
EO 7	6	114	54	162

The RIAM results indicate that positive changes occur for socio-economic components, while negative changes occur for physio-chemical and biological/ecological components in the study area. The description of positive/negative change is classified based on the ES (Table 6.7), and their range bands indicate the level of impact. In the present study, the ranges are expressed as ±A to E, but not as ±5, (The zero range is represented with 'N') because to avoid confusion between range numbers and ES values.

Table 6.6 Determination of sustainability nature based on the obtained *S* value

S level	*S* value range
Very strong sustainability	+ 0.751 to +1.000
Strong sustainability	+ 0.501 to +0.750
Weak sustainability	+ 0.251 to +0.500
Very weak sustainability	+ 0.001 to +0.250
Very weak unsustainability	0.000 to −0.250
Weak unsustainability	−0.251 to −0.500
Strong unsustainability	−0.501 to −0.750
Very strong unsustainability	−0.751 to −1.000

Based on Zand and Heir (2022)

Table 6.7 Range bands used for RIAM

Sl. No	RIAM environmental score (ES)	Range value (RV)		Description of range band
		Alphabetic	Numeric	
1	72 to 108	E	5	Major positive change/impact
2	36 to 71	D	4	Significant positive change/impact
3	19 to 35	C	3	Moderate positive change/impact
4	10 to 18	B	2	Positive change/impact
5	1 to 9	A	1	Slight positive change/impact
6	0	N	0	No change/status quo/not applicable
7	−1 to −9	−A	−1	Slight negative change/impact
8	−10 to −18	−B	−2	Negative change/impact
9	−19 to −35	−C	−3	Moderate negative change/impact
10	−36 to −71	−D	−4	Significant negative change/impact
11	−72 to −108	−E	−5	Major negative change/impact

Based on Batisha (2015)

6.2.3.1 Upstream Villages

The RIAM results show that a total of 17 components have a positive and 48 have a negative impact, while 3 have no change out of 68 environmental components due to DVC dams on upstream villages. It is also found that dams have a major positive impact on 9 socio-economic (3 socio-cultural and 6 economic/operational) components such as changes to the well-being and quality of life, quantity of recreational facilities, present and potential use of natural resources, employment opportunities, hydro-electricity generation and sales, irrigation/agricultural and industrial development. Moreover, a significant positive change has occurred in the three socio-cultural components, like cultural changes, way of life and quality of recreational facilities. No changes occur in the case of three physio-chemical components like scouring, change in humidity and bank erosion in the upstream villages. Besides, they have a major negative impact on six environmental (four physico-chemical and two biological/ecological) components such as nutrients, changes to estuaries, sedimentation, land use changes, decrease in biodiversity and impact on the threatened species. Moreover, the highest, 26 environmental (10 physio-chemical, 14 biological/ecological, and 2 socio-cultural) components have a significant negative change due to dam construction (Table 6.8 and Fig. 6.1). The physio-chemical components like heavy metals, faecal coliforms, DO, turbidity, hydraulic loss, streamflow variation, and physical composition; the biological/ecological components of flora factors such as changes in natural vegetation, disturbance of aquatic habitat and natural vegetation, decrease in biodiversity, impact of threatened species, change in species population, changes in the aquatic and mammal food web and impact on protected area, and fauna factors like migratory changes-fish and mammals, disturbance of

Table 6.8 RIAM results for upstream villages

Sl. No	RIAM environmental score (ES)	Components				Total	Class
		Physical/ chemical	Biological/ ecological	Social/ cultural	Economic/ operational		
1	72 to 108	0	0	3	6	9	E
2	36 to 71	0	0	3	0	3	D
3	19 to 35	0	0	2	0	2	C
4	10 to 18	0	0	0	0	0	B
5	1 to 9	0	0	2	1	3	A
6	0	3	0	0	0	3	N
7	−1 to −9	3	0	0	0	3	−A
8	−10 to −18	4	1	1	0	6	−B
9	−19 to −35	6	0	1	0	7	−C
10	−36 to −71	10	14	2	0	26	−D
11	−72 to −108	4	2	0	0	6	−E

Computed by the authors, 2023

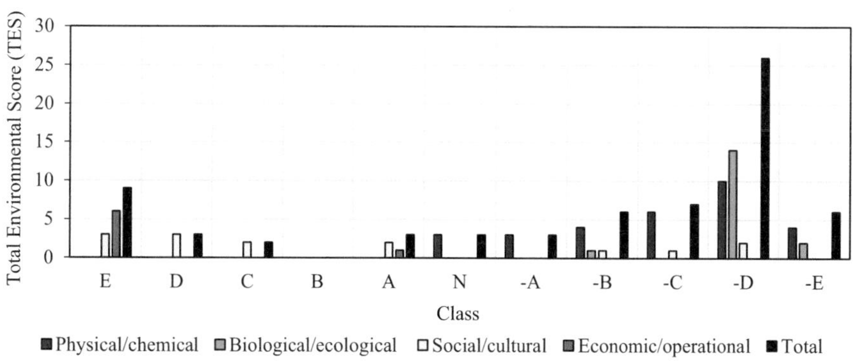

Fig. 6.1 RIAM results for the DVC project regarding upstream villages

wildlife habitat, changes in species population and changes in the mammal food web; socio-cultural components like land use change and risk of surface water contamination have a significant negative change in the upstream villages.

6.2.3.2 Downstream Villages

The RIAM results of downstream villages demonstrate that a total of 17 components have a positive and 50 have a negative impact while 1 has no change out of 68 environmental components due to DVC dams. It is also revealed that dams have a major positive impact on 8 (3 socio-cultural and 5 economic/operational) socio-economic components like cultural changes, changes to the well-being and quality of life, the present and potential use of natural resources, employment opportunities, irrigation/agricultural and industrial development. A significant positive impact of DVC dams on socio-economic components (1 socio-cultural and 2 economic/operational) like way of life, hydro-electricity sales and flood control. Besides, in the downstream villages, no changes occur in the case of 1 physio-chemical component as changes in the humidity. However, they have a major negative impact on 25 (12 physio-chemical, 11 biological/ecological, and 2 socio-cultural) environmental components. The physio-chemical components: heavy metals, faecal coliforms, BOD, COD, DO, nutrients, basin hydraulic loss, streamflow variation, changes to estuaries, sedimentation, land use changes and physical composition; the biological/ecological components of flora factors such as changes in natural vegetation, disturbance of aquatic habitat, decrease in biodiversity, the impact of threatened species, changes in species population and the aquatic food web and fauna factors like migratory changes-fish and mammals, decrease in biodiversity, impact on threatened species and changes in species population; socio-cultural components like land use change and risk of surface water contamination have a major negative impact in the downstream villages. Moreover, 14 environmental (8 physio-chemical and 6 biological/ecological) components; physio-chemical such as mercury, arsenic, lead, total organic carbon (TOC), inundation of lands, turbidity, change in pH and bank erosion and biological/ecological like disturbance of plant habitat and natural vegetation, changes in the mammal food web, impact on protected areas of flora factor and disturbance of wildlife habitat and changes in the mammal food web of fauna factor have a significant negative change due to dam construction (Table 6.9 and Fig. 6.2).

Furthermore, from the comparative study of the project impact between upstream and downstream villages, it is observed that the major positive impact of DVC dams on upstream villages is higher than the downstream villages while the major negative impact of DVC dams on upstream villages is lower than the downstream villages. Thus, these results indicate that the negative impact of the dams on downstream villages is higher than the upstream villages while the positive impact is lower than the upstream villages. The present study also reveals that the majority of physical/chemical (PC) and biological/ecological (BE) components have negatively affected by the

Table 6.9 RIAM results for downstream villages

Sl. No	RIAM environmental score (ES)	Components				Total	Class
		Physical/chemical	Biological/ecological	Social/cultural	Economic/operational		
1	72 to 108	0	0	3	5	8	E
2	36 to 71	0	0	1	2	3	D
3	19 to 35	0	0	0	0	0	C
4	10 to 18	0	0	2	0	2	B
5	1 to 9	0	0	4	0	4	A
6	0	1	0	0	0	1	N
7	−1 to −9	2	0	1	0	3	−A
8	−10 to −18	2	0	1	0	3	−B
9	−19 to −35	5	0	0	0	5	−C
10	−36 to −71	8	6	0	0	14	−D
11	−72 to −108	12	11	2	0	25	−E

Computed by the authors, 2023

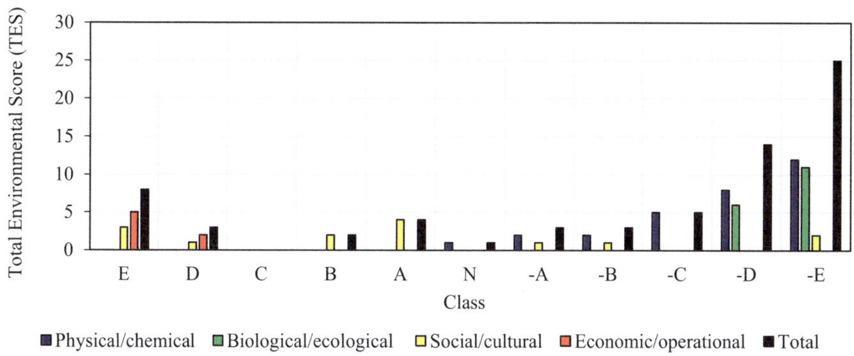

Fig. 6.2 RIAM results for the DVC project regarding downstream villages

DVC dams while the social/cultural (SC) and economic/operational (EO) components have benefited from the project in the basin area. A more or less similar finding also found by Batisha (2015), stated that major social/cultural (SC) and economic/operational (EO) components of the upstream countries benefitted from the Grand Ethiopian Renaissance Dam (GERD) project. Additionally, the major negative impacts of the project are on the water factor of physical/chemical (PC) components and on flora and fauna factors of biological/ecological (BE) in the study area. While major positive impacts of the project are on the natural resources, employment and economic development factors of economic/operational (EO) components and the quality of life factor of social/cultural (SC) components in the Damodar Valley. Hence, these results indicate that the DVC dams or DVP project have a positive impact on socio-economic components while they have a negative impact on the physical/chemical (PC) with biological/ecological (BE) components in the study area.

6.2.4 *Sustainability Model Findings*

The DVP project's sustainability has been assessed through the sustainability aspects like social, economic and environmental. In this study, a mathematical sustainability model has been applied based on the calculated RIAM results for measuring the project sustainability, i.e., whether the project is potentially sustainable or unsustainable and determining the sustainability level and nature. The project's sustainability has been evaluated based on relative ES values obtained from the original ES values of RIAM. The details of the sustainability assessment results are represented in Table 6.10. These results indicate that both upstream and downstream villages are found to have very weak unsustainability based on the determining S values (Table 6.6). These results indicate that the existing situation of the Damodar Valley is unsustainable due to the present environmental condition.

6.2.4.1 Upstream Villages

The upstream villages are found to be unsustainability as the E value (0.315) is lower than the HNI value (0.326). The calculated S value of upstream villages is −0.011, which lies in the very weak unsustainability category based on its categorical classification (Table 6.10). The reason behind the very weak unsustainability in the upstream villages is the significant impact of the project on the environment due to the creation of reservoirs and change in the water system with the pollution which negatively impacts on the PC and BE components such as nutrients, heavy metals pollution, sedimentation, land use changes, disturbance of aquatic habitat and

Table 6.10 Summary of relative ES value and sustainability assessment of the DVC project

Indices	Upstream	Downstream
Relative ES		
∑PC	2203	1823
∑PCmax	6480	6480
∑BE	996	705
∑BEmax	3672	3672
∑SC	1809	1686
∑SCmax	3024	3024
∑EO	1248	1269
∑EOmax	1512	1512
E	0.315	0.249
HNI	0.326	0.349
Sustainability/unsustainability		
S value	−0.011	−0.100
S level	Very weak unsustainability	Very weak unsustainability

Computed by the authors, 2023

natural vegetation, change in species population, migratory changes-fish and mammals, disturbance of wildlife habitat and impact on the threatened species etc.

6.2.4.2 Downstream Villages

The downstream villages are also found to be unsustainability as the *E* value (0.249) is lower than the HNI value (0.349). The calculated S value of downstream villages is −0.100 which indicates very weak unsustainability according to the categorical classification of *S* value (Table 6.10). The main reason behind the very weak unsustainability in the downstream villages also is the major negative impact of the project on the environment due to interruption and diversion of flow with the pollution which impacts the PC and BE components such as heavy metals pollution, faecal coliforms, BOD, COD, DO, nutrients, basin hydraulic loss, streamflow variation, changes to estuaries, sedimentation, land use changes, physical composition, disturbance of aquatic habitat, decrease in biodiversity, the impact of threatened species, changes in species population, and the aquatic food web, etc.

Furthermore, since the comparative analysis of sustainability between upstream and downstream villages, it is found that the sustainability in upstream villages is comparatively higher than the downstream villages. However, both lie in very weak unsustainability. The major reason behind the very weak unsustainability in the valley region is the consequent negative impacts of DVC dams on physical/chemical (PC) and biological/ecological (BE) components such as heavy metals pollution, faecal coliforms, BOD, COD, DO, nutrients, basin hydraulic loss, streamflow variation, sedimentation, physical composition, land use changes, changes to estuaries, decrease in biodiversity, disturbance of aquatic habitat and natural vegetation, change in species population and the aquatic food web, migratory changes-fish and mammals, disturbance of wildlife habitat, and impact on the threatened species. Besides, socio-cultural components like land use change and risk from surface water contamination also lead to very weak unsustainability in the basin area. The study finding is also supported by the work of Zand and Heir (2022).

6.3 Critical Outlook on Dams and Environmental Sustainability

6.3.1 RIAM and Nature of Sustainability

The impact assessment of a particular project like developmental projects including a dam on the social, economic and environmental is very crucial for sustainable development with environmental resilience and its sustainability for long-term use or continuing the project. The impact and sustainability assessment of a project is very significant for better use of natural resources and minimizing its negative impacts. A dam-like massive structural project's impact with its sustainability assessment is vital as a dam leads to socio-economic development by serving its beneficiary on one side but on the other side, it has a negative impact on the environment all over the world. In recent decades, all over the world, especially in arid and semi-arid regions, a dam has been installed on the river systems and used as a water resource and developmental tool without considering the negative consequences of it on the environment. Similarly, in India, DVC dams have been installed on the Damodar River systems and the DVC authority has a great manifesto regarding socio-economic development in the basin area leads by the DVC dams as serving their beneficiary sides such as controlling flood, providing water for industry, irrigation and domestic use, generating of hydro-electricity and creation of fishing grounds and recreation areas etc. However, they have significant negative consequences on the environment as well as river systems. The eco-geomorphological behaviour of the river systems has changed due to diversion and interruption of flow with elevated pollution that negatively impacts the riverine ecology and biota. Hence, the impact of DVC dams on socio-economic development and environment has been analysed through RIAM and its sustainability has been assessed by using a sustainability model with the help of RIAM analysis. The RIAM portrays that DVC leads to socio-economic development in the basin area (both up and downstream villages) due to significant positive consequences of SC and EO components. It is also observed that the positive consequences of the DVC dams on upstream villages are comparatively higher than the downstream villages due to the creation of reservoirs in the upstream area, water being provided for industry, irrigation and domestic use, and the creation of fishing grounds and recreation areas. The cropping pattern and intensity with crop productivity have significantly changed due to providing water in the agricultural field, especially in lean periods (arid seasons). The industrial development has occurred as DVC meet a huge industrial water requirement. They also meet a huge requirement of water for domestic use. The creation of reservoirs also leads to fishing activities and recreation facilities. Besides, DVC dams play a significant role in controlling floods in the basin area, especially the lower Damodar by storing the monsoonal flood water. These factors play a significant role in socio-economic development in the basin area. However, the RIAM reveals that DVC dams have a consequent negative impact on the environment in the Damodar Valley region (both up and downstream villages) due to significant negative consequences of PC and BE components. The consequent negative impact of the project on downstream villages is comparatively higher than the upstream villages. The prime reasons behind the consequent negative impact on the PC and BE components are the diversion and interruption of the river flow regime by the DVC dams and severe pollution. The river became a controlled river as its flow is fully controlled by the DVC dams and the river became a 'river of agony' due to environmental degradation. On the other hand, pollution is one of the significant factors for degrading the river as well as the basin environment, as the basin is home to heavy industries and mining activities with a dense population. The numerous pollutants with intoxicants from point and non-point sources such as industry, mining, agricultural fields and urban areas discharge into the river systems that impact the aquatic ecology and biota and livelihoods of the people who depend on the healthy river ecosystems. These two factors, i.e. creation of reservoirs and pollution change the river water systems which negatively impact on the PC components including air, water, climate, soil, and landscape and BE components of flora and fauna in the basin area.

Besides, from the sustainability model analysis of the project, it is observed that a very weak unsustainability has occurred in the basin area (both up and downstream villages). The reason behind the very weak unsustainability in the valley region is due to the consequent negative impacts of DVC dams on PC and BE components such as changes to estuaries, land use changes, changes in the aquatic food web, migratory changes -fish and mammals, decrease in biodiversity and impact on the threatened species. Moreover, unsustainability in the downstream villages is higher compared to upstream villages due to the negative consequences of the project on the PC and BE components, which are high at the downstream villages. This also indicated that sustainability in upstream villages is higher than in the downstream villages as a result of the positive impacts of the project on the SC and EO components being high at upstream villages. These indicate that the present situation of the Damodar Valley is unsustainable due to the existing environmental conditions.

6.3.2 Resource Management Changes in the Ecosystem

Resource management changes in ecosystems, driven by human activities and climate change, alter soil and water quality, resulting in substantial effects, including habitat loss and biodiversity decline, as well as an increased frequency of natural disasters (Munson et al., 2024; Schuurman et al., 2022). The major drivers of ecosystem change are the LULC and climate change, overexploitation and consumption of natural resources, pollution from urban-industrial, agriculture with mining, growth of non-native species and infrastructure development like dams, barrages and roads. Deforestation, expansion of built-up and agricultural areas, and mining activities decrease forest cover, fragment habitat, and reduce ecosystem services such as soil health and carbon sequestration (Safford & Vallejo, 2019; Schirpke et al., 2023). Increasing temperatures, drought, and changing weather patterns due to climate change lead to plant death, biodiversity loss, and the rising frequency and intensity of wildfires (Munson et al., 2024). Overexploitation and consumption of natural resources and pollution from point and non-point sources degrade ecosystems and their capability to provide essential services (Quadroni et al., 2024; Schuurman et al., 2022). The growth of non-native species outcompetes native species, which disrupts the ecological balance and leads to loss of biodiversity (Munson et al., 2024). Infrastructure development, like dams, barrages and roads, etc., can fragment natural habitats and promote ecosystem degradation (Hoque et al.,2025a, b).

These changes have an impact on the ecosystem, including loss of habitat and decline in biodiversity, changes in ecosystem services, rising natural disturbances and moving towards new ecological paths (Islam et al., 2025). Anthropogenic activities directly destroy or degrade habitats, which leads to substantial loss of native species and genetic diversity (Dhakal & Kattel, 2019; Safford & Vallejo, 2019). Ecosystems provide less essential services like fertile soil and clean water, but other factors, such as climate regulation, are negatively affected by human-driven alterations (Schirpke et al., 2023; Dhakal & Kattel, 2019). Natural disturbances such as drought and wildfires, which are more frequent and intense now, make ecosystems more vulnerable (Safford & Vallejo, 2019; Munson et al., 2024). Several ecosystems are evolving along new ecological pathways that differ from their historical environments, making traditional management methods insufficient (Schuurman et al., 2022).

Changing resource management in the DRB, the water resource management tools like dams and barrages of the DVC started in the mid-twentieth century (Bhattacharyya, 2011). They mainly engaged in controlling floodwaters for irrigation and hydro-electricity and providing water for industries and domestic use, which knowingly transformed the river's natural flow regime, leading to negative impact, including riverbed degradation, changed fluvial morphology, and impacts on ecosystems (Bhattacharyya, 2011; Ghosh et al., 2022; Hoque et al., 2022a). These interventions, along with urban-industrial and agricultural expansion and mining activities, have resulted in degradation of water quality and caused ecological stress through extensive pollution from sewage, urban-industrial effluents, agricultural and mining runoff (Hoque et al., 2022a). These alterations lead to a reduction in riverine habitat and threaten the sustainable development of the basin (Hoque et al. 2025a, b).

DVC's dams and barrages installed on the Damodar River and its tributaries have significantly altered the river's natural flow (Hoque et al., 2022a). This results in a reduction in seasonal flow variability and a change from a dynamic river system to a more regulated system (Ghosh et al., 2022). The river flow is controlled by DVC, and water release from dams increases the river's erosive power, leading to bed erosion and downstream degradation (Islam et al., 2023; Bhattacharyya, 2011). These interferences interrupt natural riverine ecosystems, reduce ecological function, biodiversity and generate new, less sustainable habitats downstream of dams (Ghosh et al., 2022). Besides, increasing water availability for agriculture leads to notable land use changes with agricultural development in floodplain areas. Moreover, overexploitation of the river water for urban-industrial, agricultural and domestic use also impacts its quality and quantity, leading to consequences on river ecology and biota (Hoque & Islam, 2024). The basin is a home of mineral resources, coal mining, industrial activities and a heavy population (Hoque et al., 2024). As a result, the river faces substantial pollution from untreated sewage, chemical-laden industrial effluents and mining waste with urban-industrial disposal. Urban expansion and industrial growth have resulted in severe pollution and loss of riverine habitat, further stressing the ecosystem (Hoque et al., 2025a, b). Modern agricultural practices lead to an increase in non-point sources of pollution from pesticides and fertilizers, contributing to existing environmental degradation (Hoque et al., 2023). The collective effects of dams, contamination and enormous development lead to substantial environmental stress on the Damodar River systems (Hoque et al., 2022a).

This transition is moving ecosystems away from historical norms, requiring a shift towards adaptive, ecosystem-based strategies or management approaches to accommodate environmental change and complex ecological interactions. There is an essential requirement to restore river environments by implementing environmental or ecological flow regulations and addressing the main causes of river degradation to sustain riverine biota as well as humans. Sustainable basin management needs a comprehensive approach or management systems, including controlling and managing urban-industrial and domestic effluents with their disposal, cleaning the river, releasing flow for riverine ecology and biota, community participation and policy change, to accommodate environmental flow necessities.

6.3.3 *Impacts of Anthropogenic Activities and Development on Groundwater*

Anthropogenic activities associated with development, such as industrialisation, agriculture, and urbanization, affect groundwater (GW) through pollution, overextraction, excessive erosion and land subsidence (Srivastava et al., 2025). This negative impact on groundwater threatens its quality and quantity to human and ecosystem requirements (Chakraborty et al., 2022). Besides, climate change with human activities alters recharge and depletion of groundwater, leading to stress on groundwater (Chi et al., 2022). Intensive agricultural practice, especially deep tube well irrigation, significantly reduces groundwater levels (Srivastava et al., 2025). Rapid population growth coupled with urbanisation accelerates water demand, which in turn increases groundwater extraction, leading to increased depletion (Chakraborty et al., 2022). Long-term overexploitation of groundwater for irrigation and domestic use leads to local groundwater depletion (Chi et al., 2022). Moreover, industrial activities and urban wastewater pollution leach injurious chemicals and contaminants into groundwater (Chakraborty et al., 2022). Extensive use of chemical pesticides, fertilizers, and herbicides in agricultural fields may leach into the soil, contaminating groundwater. Furthermore, anthropogenically induced climate change may alter precipitation patterns, temperature and snowmelt, which impact the groundwater recharge rate (du Plessis, 2018; Kalantari et al., 2019). The combined effect of rising water demand and potential alteration in recharge as a result of climate change could deteriorate water scarcity, leading to an imbalance between water demand and supply (Tan et al., 2025). Groundwater extraction can affect surface water flow, changing stream temperature, nutrient concentrations, and oxygen levels, harming aquatic or river ecosystems and biota. Prolonged groundwater depletion may also lead to land subsidence, where the surface sinks as the water is removed from underground aquifers (Ferng, 2007; Chi et al., 2022).

Anthropogenic activities and development in the DRB involving industrialisation, urbanisation, agricultural expansion and mining (coal, sand and other minerals) have notably changed the Damodar River hydrology and water quality, resulting in rising groundwater pollution and degrading the river ecosystems (Chakraborty et al., 2022; Hoque et al., 2022a). However, the creation of reservoirs in the basin area by DVC also promotes local groundwater recharge. Industrialisation and urbanisation lead to substantial contamination concentration in the river systems and groundwater. Besides, the extensive extraction of coal and minerals in the basin has amplified the net contamination load on both surface water and groundwater (Ambade et al., 2021). Intensive agricultural practices also lead to significant pollution concentrations in river systems and groundwater. Intensive agriculture in the basin also leads to groundwater depletion due to the overexploitation of groundwater resources. Thus, contamination from coal and other mineral mines and urban-industrial effluents with disposal, and agricultural runoff, directly impacts groundwater, resulting in high concentrations of contaminants (Maity et al., 2025; Ambade et al., 2021). Elevation of heavy metals in groundwater poses human health risks (non-carcinogenic and carcinogenic) to residents of Santuri block, Purulia district, West Bengal, especially to children (Chakraborty et al., 2022).

Thus, the implementation of strict policies and regulatory mechanisms that can control groundwater extraction and prevent its overexploitation (Upendra et al., 2024; Srivastava et al., 2025). Promotion of water conservation practices in irrigation and other sectors involving water recycling and reuse, and implementation of water assessments, is essentially required to decrease pressure on groundwater. Besides, understanding the complexity of aquifers through comprehensive measurement can deliver a theoretical foundation for sustainable groundwater management (Chi et al., 2022; Srivastava et al., 2025). Empowering local communities to manage groundwater could encourage sustainable management practices and ensure prolonged accessibility (Wali et al., 2024).

6.3.4 Impacts of Anthropogenic Activities and Development on Soil Water

Human activities such as agriculture, industrialisation, urbanisation and mining damage the quality and quantity of soil water by presenting contaminants (nutrients, heavy metals, and chemicals). Deforestation and hardscaping increase soil erosion and runoff, and alter the natural water balance. These alterations lead to soil degradation, salinisation, decreased groundwater recharge, biodiversity loss, and possible human health risks (Gavrilaş et al., 2025). Intensive agriculture leads to surplus pesticides and fertilizers, whereas urban-industrial effluents and mining activities lead to heavy metals and other injurious chemical concentrations in soil and water systems (Khatri & Tyagi, 2015). Rising nutrients, such as nitrogen and phosphorus from agricultural effluents, can lead to increased eutrophication levels in aquatic ecosystems and disturb the natural balance (Zhu et al., 2022). Extreme irrigation and canal drainage increase soil water salinity, and the land becomes unproductive (Patel et al., 2020). Urban-industrial effluents and wastewater irrigation introduce pathogenic bacteria and increase bacterial pollution in soil and groundwater (Chapelle, 2000; Asirifi et al., 2021). Clearing vegetation and the removal of surface covering due to deforestation and built-up areas lead to a decrease in the ability of land to absorb and hold water, which reduces groundwater recharge (Kumar et al., 2022). Besides, extensive agriculture, deforestation and mining increase soil erosion and accelerate surface runoff, changing natural water flow and storage. Hence, LULC changes crucially disturb the local and regional hydrological cycle, which impacts baseflow and surface and dry-season runoff. These lead to soil erosion, loss of surface soil fertility, and rising contamination, reducing soil fertility and damaging ecosystems and biodiversity (Kumar et al., 2022). Additionally, polluted soil water poses direct or indirect health risks to humans through contaminated food or water consumption, particularly for heavy metals and bacteria (Khatri & Tyagi, 2015; Roşca et al., 2020). Moreover, changing water quality due to pollution significantly impacts aquatic ecosystems, which leads to an effect on aquatic biodiversity and the environment.

Human activities in the form of development in the DRB, such as industrialization and coal and other minerals mining, lead to elevated soil erosion and water pollution from heavy metals and toxic contaminants, reduce soil fertility, and worsen water quality. Coal washeries and other industries' effluents containing elevated levels of heavy metals (copper, zinc, lead, iron, manganese, chromium), total suspended solids (TSS), total dissolved solids (TDS), phenols, oil, and grease, are polluting both soil and river water (Ghosh et al., 2025). Heavy metals from industrial and mining activities gather in soils as well as agricultural fields, which leads to rising salinity and poses threats to agriculture (Hoque et al., 2023). Additionally, urbanisation and modern intensive agricultural practices, including the use of fertilizers and pesticides, also lead to heavy metal concentration in soil and river water. Moreover, the installation of dams and barrages by DVC, however, useful for flood control and irrigation, notably modifies ecological or environmental flow patterns, affecting river ecosystems and biota and the overall health of the basin.

6.3.5 *Impacts of Anthropogenic Activities and Development on Soil Productivity*

Human activities such as deforestation, industrialisation, urbanisation, and intensive agricultural practice reduce soil productivity due to loss of organic matter, pollution, erosion, and compaction (Kumar et al., 2022). These anthropogenically driven processes modify soil structure, biological activity and nutrient cycling, which threaten soil fertility, crop production, and ecosystem health (Gantulga et al., 2023). These soil-landscape pattern alterations are changing soil structure and properties. The weathering and formation rates have also changed, leading to lateralization, podzolization, and soil acidity are increasing (Kumar et al., 2022). This leads to soil degradation and loss of crop yields and essential ecosystem services (Santorufo et al., 2021). Deforestation, overgrazing, and ploughing eradicate tree roots, which stabilize soil, raising the soil erosion rate by 10–100 times than natural processes (Kumar et al., 2022). Built-up and heavy agricultural machinery activities compact soil, reduce pore space, impede water and air movement, and inhibit root growth (Zhou & Gu, 2024). Intensive agricultural practice, deforestation and acid deposition reduce soil nutrients, whereas industrial activities and forestry increase soil acidity (Grieve, 2001). Industrial procedures, exhaust fumes, and waste dumping introduce heavy metals and other pollutants like microplastics into soils, leading to soil contamination and decreasing biological activity (Sarkar et al., 2022). These combined factors result in an important loss of soil fertility and a reduction in the capacity of the land to support crops and plants (Kumar et al., 2022; Gantulga et al., 2023). Besides, soil quality degradation disrupts important ecosystem services such as carbon storage, water filtration, and food and raw material supply, affecting livelihoods and leading to socio-economic instability (Jones & Rowe, 2017). Moreover, contamination of soil and water and climate change lead to mass migration and conflict over resources (Ukpere, 2024).

Anthropogenic activities in the form of development in the DRB involving industrialization, urbanization, agricultural expansion, and mining (coal, sand, and other minerals) have notably degraded soil and water quality, due to severe soil erosion, pollution, heavy sedimentation in rivers, and conversion of productive land into desert. Intensive mining leads to severe soil erosion, with parts experiencing significant soil loss, especially in the upper basin (Ghosh et al., 2025). This soil erosion has converted large areas into unproductive wastelands. Deforestation, coal, and mineral mining in the basin area lead to excessive soil erosion, resulting in the loss of productive soils and the creation of unproductive wasteland (Ghosh et al., 2025). A comprehensive basin management strategy involving integrated land and watershed management and active community participation is required to mitigate these problems and achieve sustainable development in the basin (Hoque et al., 2022a; Ghute et al., 2023).

6.4 Concluding Note

The RIAM results indicate that the positive impact of the DVC project on socio-economic components is high in upstream villages, while the negative impact of it on physio-chemical and biological/ecological components is lower than in the downstream villages. The sustainability model analysis of the project reveals a very weak unsustainability for both upstream and downstream villages in the Damodar Valley region due to the consequent negative impacts on PC and BE components from the DVC dams. Besides, resource management related to development activities affects soil and water quality, leading to habitat loss, decline in biodiversity, and increased natural disasters. Water management tools such as DVC dams and barrages alter the river's natural flow regime, causing negative impacts like riverbed degradation, changes in fluvial morphology, and ecosystem disruption. These interventions, combined with agricultural and urban-industrial expansion and mining activities, have degraded water quality and caused ecological stress due to pollution from these sectors. Furthermore, overexploitation of river water for urban, industrial, agricultural, and domestic use also impacts its quantity and quality, affecting river ecology and biota. These changes in both quality and quantity lead to habitat reduction and threaten the sustainable development of the basin.

Human activities, including industrialisation, agriculture, urbanisation, and mining, influence groundwater quantity and quality, impacting both humans and ecosystems. Additionally, climate change driven by human activity may alter precipitation patterns and temperatures, affecting groundwater recharge rates. Although reservoir creation in the basin can promote local groundwater recharge, activities like deforestation, hardscaping, and other developmental projects intro duce contaminants such as heavy metals, affecting soil and water quality. These activities also increase soil erosion and runoff, disrupting the natural balance of soil water and altering water flow and storage. They diminish soil productivity through the loss of organic matter, pollution, erosion, and compaction. Human-driven processes modify soil structure, biological activity, and nutrient cycling, threatening soil fertility, crop yields, and ecosystem health. Soil degradation also hampers essential ecosystem services like carbon storage, water filtration, and the supply of food and raw materials, thereby affecting livelihoods and causing socio-economic instability. Furthermore, soil and water contamination alongside climate change can lead to mass migration and conflicts over resources. Deforestation, coal and mineral mining within the basin cause excessive soil erosion, resulting in the loss of productive land and the formation of unproductive wastelands. In summary, while development-related activities bring socio-economic benefits, they also harm the basin's ecosystems, groundwater, soil water, and soil productivity. The analysis retained in this chapter and excerpts from the previous chapters (Chaps. 2–6) clearly indicate the unsustainability of the fluvial system and adjacent villages with elevated pollution for ecology, irrigation, and human health that need some management measures for the sustainable restoration of the environmental flow. This has been taken up in the next chapter, i.e., Chap. 7.

References

Altinbilek, D. (2002). The role of dams in development. *Water Science and Technology, 45*(8), 169–180.

Ambade, B., Sethi, S. S., Kurwadkar, S., Kumar, A., & Sankar, T. K. (2021). Toxicity and health risk assessment of polycyclic aromatic hydrocarbons in surface water, sediments and groundwater vulnerability in Damodar River Basin. *Groundwater for Sustainable Development, 13*, 100553. https://doi.org/10.1016/j.gsd.2021.100553

Asirifi, I., Kaetzl, K., Werner, S., Saba, C. K., Abagale, F. K., Amoah, P., & Marschner, B. (2021). Pathogen and heavy metal contamination in urban agroecosystems of northern Ghana:

Influence of biochar application and wastewater irrigation. *Journal of Environmental Quality, 50*(5), 1097–1109. https://doi.org/10.1002/jeq2.20260
Batisha, A. F. (2015). Sustainability assessment in transboundary context: Grand Ethiopian Renaissance dam. *Modeling Earth Systems and Environment, 1*(4), 36. https://doi.org/10.1007/s40808-015-0033-2
Bhattacharyya, K. (2011). *The Lower Damodar River, India: Understanding the human role in changing fluvial environment* (p. 295). Springer. https://doi.org/10.1007/978-94-007-0467-1
Chakraborty, B., Roy, S., Bera, B., Adhikary, P. P., Bhattacharjee, S., Sengupta, D., & Shit, P. K. (2022). Evaluation of groundwater quality and its impact on human health: A case study from Chotanagpur plateau fringe region in India. *Applied Water Science, 12*(3), 25. https://doi.org/10.1007/s13201-021-01539-6
Chapelle, F. H. (2000). *Ground-water microbiology and geochemistry*. John Wiley & Sons.
Chi, G., Su, X., Lyu, H., Li, H., Xu, G., & Zhang, Y. (2022). Prediction and evaluation of groundwater level changes in an over-exploited area of the Baiyangdian Lake Basin, China under the combined influence of climate change and ecological water recharge. *Environmental Research, 212*, 113104. https://doi.org/10.1016/j.envres.2022.113104
Damodar Valley Corporation (2021). *DVC Reservoir Overview*. www.dvc.gov.in/dvcwebsite_newl/damsbarrages/DamodarValleyCoporation
Dhakal, B., & Kattel, R. R. (2019). Effects of global changes on ecosystems services of multiple natural resources in mountain agricultural landscapes. *Science of the Total Environment, 676*, 665–682. https://doi.org/10.1016/j.scitotenv.2019.04.276
du Plessis, A. (2018). Climate change: Current drivers, observations and impacts on the Globe's natural and human systems. In *Water as an inescapable risk: Current global water availability, quality and risks with a specific focus on South Africa* (pp. 27–53). Springer. https://doi.org/10.1007/978-3-030-03186-2_3
Ferng, J. J. (2007). Human freshwater demand for economic activity and ecosystems in Taiwan. *Environmental Management, 40*(6), 913–925. https://doi.org/10.1007/s00267-007-9023-0
Gantulga, N., Iimaa, T., Batmunkh, M., Surenjav, U., Tserennadmin, E., Turmunkh, T., et al. (2023). Impacts of natural and anthropogenic factors on soil erosion. *Proceedings of the Mongolian Academy of Sciences*, 3–18. https://doi.org/10.5564/pmas.v63i02.1416
Gavrilaş, S., Burescu, F. L., Chereji, B. D., & Munteanu, F. D. (2025). The impact of anthropogenic activities on the catchment's water quality parameters. *Water, 17*(12), 1791. https://doi.org/10.3390/w17121791
Ghosh, S., & Mistri, B. (2013). Performance of DVC in flood moderation of lower Damodar River India and emergent risk of flood. *Eastern Geographer, 19*(1), 55–66.
Ghosh, S., Hoque, M. M., Saha, U. D., & Islam, A. (2022). Assessment of dam-induced changes in ecogeomorphological behaviour and fluvial functionality in the Damodar River, West Bengal, India. *AQUA—Water Infrastructure, Ecosystems and Society, 71*(6), 722–750. https://doi.org/10.2166/aqua.2022.003
Ghosh, J., Das, S., Ghosh, S., Midya, K., Das, S., & Mishra, V. N. (2025). Soil erosion vulnerability analysis of Damodar River Basin, India using Revised Universal Soil Loss Equation (RUSLE) in Google Earth Engine (GEE). *Discover Geoscience, 3*(1), 1–31. https://doi.org/10.1007/s44288-025-00191-w
Ghute, B. B., Shaikh, M. B., & Halder, B. (2023). Impact assessment of natural and anthropogenic activities using remote sensing and GIS techniques in the Upper Purna River basin, Maharashtra, India. *Modeling Earth Systems and Environment, 9*(2), 1507–1522. https://doi.org/10.1007/s40808-022-01576-3
Grieve, I. C. (2001). Human impacts on soil properties and their implications for the sensitivity of soil systems in Scotland. *Catena, 42*(2–4), 361–374. https://doi.org/10.1016/S0341-8162(00)00147-8
Hoque, M. M., & Islam, A. (2024). Spatio-temporal dynamics of ecological, bacteriological, and overall water quality of the Damodar River, India. *Environmental Science and Pollution Research, 31*(12), 18465–18484. https://doi.org/10.1007/s11356-024-32185-5

Hoque, M. M., Islam, A., & Ghosh, S. (2022a). Environmental flow in the context of dams and development with special reference to the Damodar Valley Project, India: A review. *Sustainable Water Resources Management, 8*(3), 62. https://doi.org/10.1007/s40899-022-00646-9

Hoque, M. M., Islam, A., & Mahammad, S. (2022b). Assessing the surface and bottom river water quality for drinking purpose and human health risk analysis: a study of Damodar River, India. *Arabian Journal of Geosciences, 15*(23), 1734. https://doi.org/10.1007/s12517-022-10987-6

Hoque, M. M., Islam, A., Islam, A. R. M. T., Pal, S. C., Mahammad, S., & Alam, E. (2023). Assessment of soil heavy metal pollution and associated ecological risk of agriculture dominated mid-channel bars in a subtropical river basin. *Scientific Reports, 13*(1), 11104. https://doi.org/10.1038/s41598-023-38058-0

Hoque, M. M., Islam, A., Sarkar, B., & Saha, U. D. (2024). Assessing the surface and bottom river water quality for irrigation: A study of Damodar River, India. *International Journal of Energy and Water Resources, 8*(4), 545–562. https://doi.org/10.1007/s42108-022-00206-z

Hoque, M. M., Islam, A., Pal, S. C., Sarkar, B., & Hossain, Z. M. (2025a). Assessment of fish habitat suitability in the wake of elevated pollution of a tropical river in India. *Environmental Science and Pollution Research*, 1–15. https://doi.org/10.1007/s11356-024-35685-6

Hoque, M. M., Pal, S. C., Mandal, S., & Islam, A. (2025b). Assessment of mollusc habitat suitability of an anthropogenically altered tropical river in India. *Environmental Monitoring and Assessment, 197*(10), 1–18. https://doi.org/10.1007/s10661-025-14593-3

Islam, A., Sardar, N., Mohinuddin, S., Hoque, M. M., Sengupta, S., Das, B. C., ... & Sengupta, B. (2023). Quasi-equilibrium channel metamorphosis in planform of a subtropical river in India in post-dam period. *Catena, 221*, 106793. https://doi.org/10.1016/j.catena.2022.106793

Islam, A., Das, B. C., Mahammad, S., Hoque, M. M., Pham, Q. B., Sarkar, B., ... & Barman, S. D. (2025). Assessing river water quality for ecological risk in the context of a decaying river in India. *Environmental Science and Pollution Research, 32*(16), 10223–10239. https://doi.org/10.1007/s11356-024-33684-1

Jones, D. L., & Rowe, E. C. (2017). Land reclamation and remediation, principles and practice. In *Encyclopedia of applied plant sciences* (pp. 304–310). Elsevier Ltd. https://doi.org/10.1016/B978-0-12-394807-6.00014-9

Kalantari, Z., Ferreira, C. S. S., Page, J., Goldenberg, R., Olsson, J., & Destouni, G. (2019). Meeting sustainable development challenges in growing cities: Coupled social-ecological systems modeling of land use and water changes. *Journal of Environmental Management, 245*, 471–480. https://doi.org/10.1016/j.jenvman.2019.05.086

Khatri, N., & Tyagi, S. (2015). Influences of natural and anthropogenic factors on surface and groundwater quality in rural and urban areas. *Frontiers in Life Science, 8*(1), 23–39. https://doi.org/10.1080/21553769.2014.933716

Kumar, G., Baweja, P., & Gandhi, P. B. (2022). Impact of anthropogenic activities on soil patterns and diversity. In *Structure and functions of pedosphere* (pp. 319–337). Springer. https://doi.org/10.1007/978-981-16-8770-9_13

Loomis, J. (2002). Quantifying recreation use values from removing dams and restoring free-flowing rivers: A contingent behavior travel cost demand model for the Lower Snake River. *Water Resources Research, 38*(6), 1–8.

Maity, S., Maiti, R., & Mukherjee, S. (2025). Catchment scale land use and hydrogeomorphology relationship in assessing river water quality using SWAT and RDA modelling in the Damodar River basin, West Bengal, India. *Advances in Space Research*. https://doi.org/10.1016/j.asr.2025.03.036

Munson, S. M., Vaughn, A. L., Petersen, B., Bradford, J. B., & Duniway, M. C. (2024). Natural resource management confronts the growing scale and severity of ecosystem responses to drought and wildfire. *Ecology and Society, 29*(4). https://doi.org/10.5751/ES-15517-290417

Pastakia, C. M., & Jensen, A. (1998). The rapid impact assessment matrix (RIAM) for EIA. *Environmental iew, 18*(5), 461–482. https://doi.org/10.1016/S0195-9255(98)00018-3

Patel, M. P., Gami, B., Patel, A., Patel, P., & Patel, B. (2020). Climatic and anthropogenic impact on groundwater quality of agriculture dominated areas of southern and central Gujarat,

India. *Groundwater for Sustainable Development, 10*, 100306. https://doi.org/10.1016/j.gsd.2019.100306

Phillips, J. (2009). *The development and application of a geocybernetic model of sustainability* (Doctoral dissertation, Exeter University).

Quadroni, S., Cesarini, G., De Santis, V., & Galafassi, S. (2024). Interconnected impacts of water resource management and climate change on microplastic pollution and riverine biocoenosis: A review by freshwater ecologists. *Journal of Environmental Management, 372*, 123363. https://doi.org/10.1016/j.jenvman.2024.123363

Roşca, O. M., Dippong, T., Marian, M., Mihali, C., Mihalescu, L., Hoaghia, M. A., & Jelea, M. (2020). Impact of anthropogenic activities on water quality parameters of glacial lakes from Rodnei mountains, Romania. *Environmental Research, 182*, 109136. https://doi.org/10.1016/j.envres.2020.109136

Safford, H. D., & Vallejo, V. R. (2019). Ecosystem management and ecological restoration in the Anthropocene: Integrating global change, soils, and disturbance in boreal and Mediterranean forests. In *Developments in soil science* (Vol. 36, pp. 259–308). Elsevier.

Santorufo, L., Memoli, V., Panico, S. C., Esposito, F., Vitale, L., Di Natale, G., et al. (2021). Impact of anthropic activities on soil quality under different land uses. *International Journal of Environmental Research and Public Health, 18*(16), 8423. https://doi.org/10.3390/ijerph18168423

Sarkar, A., Deb, S., Ghosh, S., Mandal, S., Quazi, S. A., Kushwaha, A., et al. (2022). Impact of anthropogenic pollution on soil properties in and around a town in Eastern India. *Geoderma Regional, 28*, e00462. https://doi.org/10.1016/j.geodrs.2021.e00462

Schirpke, U., Tasser, E., Borsky, S., Braun, M., Eitzinger, J., Gaube, V., et al. (2023). Past and future impacts of land-use changes on ecosystem services in Austria. *Journal of Environmental Management, 345*, 118728. https://doi.org/10.1016/j.jenvman.2023.118728

Schuurman, G. W., Cole, D. N., Cravens, A. E., Covington, S., Crausbay, S. D., Hoffman, C. H., et al. (2022). Navigating ecological transformation: Resist–accept–direct as a path to a new resource management paradigm. *BioScience, 72*(1), 16–29. https://doi.org/10.1093/biosci/biab067

Srivastava, S., Singh, A., & Dwivedi, D. (2025). Facing groundwater depletion in India: The role of human activities and climate extremes★. *Groundwater for Sustainable Development*, 101430. https://doi.org/10.1016/j.gsd.2025.101430

Tan, X., Liu, B., Tan, X., Huang, Z., & Fu, J. (2025). Combined impacts of climate change and human activities on blue and green water resources in a high-intensity development watershed. *Hydrology and Earth System Sciences, 29*(2), 427–445. https://doi.org/10.5194/hess-29-427-2025

Tortajada, C. (2015). Dams: An essential component of development. *Journal of Hydrologic Engineering, 20*(1), A4014005. https://doi.org/10.1061/(ASCE)HE.1943-5584.0000919

Ukpere, D. T. P (2024). *Anthropogenic activities, soil quality and crop production in Ukwa East area of Abia State, Nigeria*. https://doi.org/10.56201/ijaes.v10.no7.2024.pg144.155

Upendra, B., Nanda, K., Jesuraja, K., Ciba, M., Sreenivasulu, G., & Anoop Krishnan, K. (2024). Analysis of human activities, built-up and rainfall impacts on groundwater potential zones using GIS and AHP techniques: A case of Karaman River, southern western ghats, India. *Water Conservation Science and Engineering, 9*(2), 61. https://doi.org/10.1007/s41101-024-00294-4

Wali, S. U., Usman, A. A., Usman, A. B., Abdullahi, U., Mohammed, I. U., & Hayatu, J. M. (2024). Impact of geology on hydrogeological and hydrochemical characteristics of groundwater in tropical environments: A narrative review. *IJH, 8*(6), 202–221. https://doi.org/10.15406/ijh.2024.08.00392

WCD. (2000). *Dams and development: A new framework for decisionmaking*. Earthscan Publications Ltd.

Zand, A. D., & Heir, A. V. (2022). Integration of rapid impact assessment matrix method and sustainability modeling for management of municipal solid waste transfer stations in cold regions. *Modeling Earth Systems and Environment*, 1–12. https://doi.org/10.1007/s40808-021-01248-8

Zhou, Y., & Gu, B. (2024). The impacts of human activities on Earth Critical Zone. *Earth Critical Zone, 1*(1), 100004. https://doi.org/10.1016/j.ecz.2024.100004

Zhu, X., Wang, L., Zhang, X., He, M., Wang, D., Ren, Y., et al. (2022). Effects of different types of anthropogenic disturbances and natural wetlands on water quality and microbial communities in a typical black-odor river. *Ecological Indicators, 136*, 108613. https://doi.org/10.1016/j.ecolind.2022.108613

Chapter 7
Environmental Flow Management

7.1 Concept of Environmental Flow Restoration

Environmental flow (E. flow) restoration refers to the practice of managing and restoring natural water flows in rivers, streams, and other water bodies to maintain the health of aquatic ecosystems and the services they provide (Hoque et al., 2022a). This concept recognizes that water is not just a resource for human use but also a critical component of ecosystems that support various species and ecological processes (Hoque et al., 2025a, b). Human activities, such as dam construction, irrigation, urban development, and water extraction, often alter the natural flow patterns of rivers and disrupt the balance of aquatic ecosystems (Hoque et al., 2024; Hoque and Islam, 2024). These alterations can lead to a range of ecological problems, including habitat degradation, loss of biodiversity, water quality issues, and diminished ecosystem services (Islam et al., 2025; Hoque et al., 2024). Environmental flow restoration seeks to address these issues by restoring natural flow regimes as much as possible.

Environmental flow restoration includes a few key components (i) flow regime (the timing, frequency, duration, and magnitude of water flow in a river), (ii) ecosystem needs (different species and habitats with varying requirements for water flow), (iii) stakeholder engagement (balancing the needs of both ecosystems and human societies), (iv) science-based approaches (scientific research to determine the specific flow requirements for different ecosystems and species), (v) monitoring and adaptive management (monitoring to assess their effectiveness of the restoration plans), (vi) infrastructure modifications (physical modifications to dams, and other structures for retaining natural flow patterns), and (vii) legal and policy frameworks (supportive legal and policy frameworks for ecological purposes). Overall, environmental flow restoration represents a holistic approach to water management that seeks to strike a balance between human water needs and the preservation of healthy

M. M. Hoque, A. Islam, *Dams, Development and Environmental Flow in the Damodar River Basin*, https://doi.org/10.1007/978-3-032-11161-6_7

aquatic ecosystems. It is a critical tool in the broader effort to address water scarcity and protect the environment in the face of increasing human demands on water resources. Thus, this chapter intends to address two sub-themes—(Sect. 7.2) rationale of the environmental flow restoration and (Sect. 7.3) strategies of environmental flow restoration as discussed in the following sections.

7.2 Rationale of Environmental Flow Restorations

The restoration of environmental flow is necessarily required as it maintains and sustains freshwater and estuarine ecosystems, along with the livelihoods and well-being of people who rely on these ecosystems. Specifically, river environmental flow restoration is needed for managing riverine ecosystems (Table 7.1). It is also required for cultural and spiritual needs in the Indian context. Environmental flow encompasses a wide range of values and benefits from rivers, instead of focusing on water supply, energy, recreation, or flood control for socio-economic benefits. The structural interventions such as DVC dams and barrage with other river structures (like weirs and bridges), along with non-structural interventions alter the environmental flow of the Damodar River systems, especially downstream which impact water quality, temperature, sediment movement, and deposition, fish and wildlife, and the livelihoods of people who depend on healthy ecosystems of the river. Environmental flows seek to maintain these river functions while at the same time providing for traditional off-stream benefits.

Table 7.1 Requirements of environmental flow

Aspects	E. flow needs
Aquatic animals	Maintaining physical habitat, fish migration, spawning/egg laying, water quality suitability
Riparian vegetation	Maintaining river banks' soil moisture, seed distribution, and nutrient deposition
River sand	Sand transportation and separation
Estuaries	Maintaining freshwater/salt balance and ocean-to-estuary connection
Aquifers and groundwater	Recharging the aquifer and groundwater
Floodplains	Floods for floodplain inundation
Aesthetics	Maximizing natural aesthetic features
Recreational and cultural features	Maintaining water quality, algae flushing, and flow for sediment
Ecosystem services	Maintaining and sustaining ecosystem functions and biodiversity
Overall environmental protection	Some or all of the above

Based on Dutta et al. (2020)

7.2.1 *Environmental Flow Requirements of Indian Rivers*

Rapid urbanization, industrialization, and intensification of agriculture are very pertinent issues in India (Hoque et al., 2022b), and these anthropogenic factors together deeply affect the hydrogeomorphic and environmental structure of Indian Rivers (Islam et al., 2023), creating major environmental concerns in the Anthropocene (Hoque et al., 2022a). Hundreds of multi-purpose dams and reservoirs, like the Damodar Valley Corporation (DVC), for water supply, irrigation, hydropower, flood management, and fisheries have been developed. Besides, in the Indo-Gangetic Plains, many floodplains (viz., Ganga, Ghaggar, Kosi, Yamuna, Bhagirathi-Hooghly Rivers, etc.) have been detached from the main rivers by elevated embankments, and the remaining riparian lands are under intensive agriculture, urbanization, and grazing pressure (Smakhtin & Anputhas, 2006). Human settlements and anthropogenic activities (like mining, damming, deforestation, and pollution generation) have degraded the environmental quality of river valleys and increased sedimentation in all rivers of India (Hoque et al., 2023). All the affected riverine biota and species composition have been altered, and many species have nearly vanished or are endangered (like Fishing Cat, Smooth-coated Otter, Gangetic Dolphin, Gharial, Ganges Shark, Sarus Crane, etc.). In India, the river basins, mainly the Ganga-Brahmaputra-Meghna Delta, are now strongly affected by flow fragmentation and regulation due to high demand and global climate change (Smakhtin & Anputhas, 2006). Central Pollution Control Board (CPCB) data show that the ten dirtiest river stretches in 2018 were found on river Vasishta (Tamil Nadu), Ghaggar (Haryana and Punjab), Bhadra (Gujarat), Mithi (Maharashtra), Sabarmati (Gujarat), Hindon (Uttar Pradesh), Sutlej (Punjab), Khori (Gujarat), and Thirumanimuthar (Tamil Nadu). CPCB is currently monitoring 445 rivers in India and the number of polluted stretches (in the category of 3–6 mg/L of BOD level) has increased from 36 (2009) to 175 (2018). West Bengal Pollution Control Board (WBPCB) identifies 17 stretches of river as polluted based on BOD, DO, and total coliform (Table 7.2).

The expressions such as ‘environmental flows’ and ‘water for nature’ mean an allocation of river water must be made for nature at first, then allocating water for different uses (Smakhtin & Anputhas, 2006). It is found that the status of environmental flow research is not very satisfactory in India and this research is now under a growing stage of development. The National Commission for Integrated Water Resource Development Plan (NCIWRDP) efficiently established that it was not achievable to calculate approximately the amount of water required for different environmental purposes (Smakhtin & Anputhas, 2006). The Supreme Court of India had made a judgement on the issue of minimum flow. In 1999, the Supreme Court of India directed the central government to guarantee a minimum flow of 10 m^3/s in the Yamuna River as it flows through New Delhi to improve its water quality (Smakhtin & Anputhas, 2006). The Environmental flow of the Baitarani River has been estimated based on the need to sustain 7-day minimum and 1-day maximum channel flows in the river. The entire country has an environmental flow of about

Table 7.2 Polluted stretches of rivers need environmental flow in West Bengal

Sl. No.	River	Polluted stretch	BOD when identified as polluted (mg/L)	BOD at present	DO at present	Total coliform at present (MPN/100 mL)	Fitness comment Class D
1	Vindyadhari	Haroa bridge to Manlancha burning Ghat	26.7–45.0	14.69	1.8	130,000	Not fit for bathing
2	Mahananda	Siliguri to Binaguri	6.5–25	21.0	4.8	220,000	Not fit for bathing
3	Churni	Santipur town to Majhadia	10.3–11.3	3.1	4.0	35,000	Not fit for bathing
4	Dwarka	Tarapith to Sadhak Bamdeb Ghat	5.6–17	3.15	7.3	54,000	Not fit for bathing
5	Bhagirathi-Hooghly	Tribeni to Diamond Harbour	5.0–12.2	4.5	7.1	170,000	Not fit for bathing
6	Damodar	Durgachakm to Dishergarh	4.4–8.2	3.95	7.1	12,000	Not fit for bathing
7	Jalangi	Lal Dighi to Krishnanagar	8.3	3.8	6.5	17,000	Not fit for bathing

Source: WBPCB (2020)

476 km^3, which constitutes approximately 25% of total renewable water resources in India (Smakhtin & Anputhas, 2006).

7.2.2 *The Environmental Flow of the Damodar River*

7.2.2.1 River of Agony

The Damodar River plays a key role in providing a source of drinking water, water for irrigation and power, industrial needs, and coal mining activities in the basin. The river once known as the '*River of Sorrow*' because of the severity of the flood (Mahammad & Islam, 2021) has now turned into a '*River of Agony*' due to the environmental degradation of the river resulting from the indiscriminate discharge of domestic, industrial, and mining wastes from the basin (Patra, 2008). The inhabitants of the Damodar Floodplain now evidence the agony side of DVC dams. After the dam construction, each person expected that the extreme floods would completely vanish from this region. Still, the flood has now modified its character, promoting low peak but longer duration (more than 1 week), due to drainage congestion, delinking of palaeochannel from the main river, and the inability of regulated discharge due to reservoir siltation. Moreover, the river stretches about 68 km from the downstream of Panchet Dam to the upstream of Durgapur Barrage in West Bengal passing through the main industrial belt of eastern India and covering important cities like Kulti, Barakar, Asansol, Raniganj, and Durgapur which have been affected due to environmental stress and water pollution. The stretch is receiving a large number of point and non-point sources of pollution from thermal power plants, collieries and coal-based industries, steel, cement, fertilizers, chemical, and other plants as well as domestic and agricultural areas that lead to water pollution to the river either directly or through numerous drains and *nallahs* (Patra, 2008).

Environmentalists often say that the industrialists exploiting the area's resources are not willing to make a positive change in the physical environment. It was informed by public media that about 200,000 L of furnace oil spilled into the water of Damodar from the Bokaro Steel Plant (BSP) on 2nd April 1990. To check that disaster the authorities of BSP and government took at least 4 days. It was informed that the spilled oil reached up to Durgapur (150 km from BSP) after a week and five million people in this area used the contaminated water of Damodar. That disaster was not noticed by any pollution control authorities in India. That incident was first noticed by a downstream thermal power plant which abruptly found the inefficiency of the cooling process in its boilers, raising the alarm. That event of 1990 gave us a vulnerability alarm of industrial accidents in the river because more than 50 major industries and more than 400 industrial units have developed on the banks of the Damodar River. The states of Jharkhand and West Bengal depend on the river water and the power plants developed in the valley. The thermal power plants of DVC alone generate about 1800 MW of electricity. Considering the environmental concern, the biggest threat to the Damodar River is the vast mineral resources of the Chotanagpur Plateau which are intensively exploited due to the mounting demand of the Indian economy. The River is badly affected both physically and ecologically by dumping urban waste material with extensively industrial effluent discharge and siltation reducing river water flow (Fig. 7.1). The River course is also influenced by changing land use and cropping pattern, creating fluvial metamorphosis. Thus, the river tends to become an ecological desert.

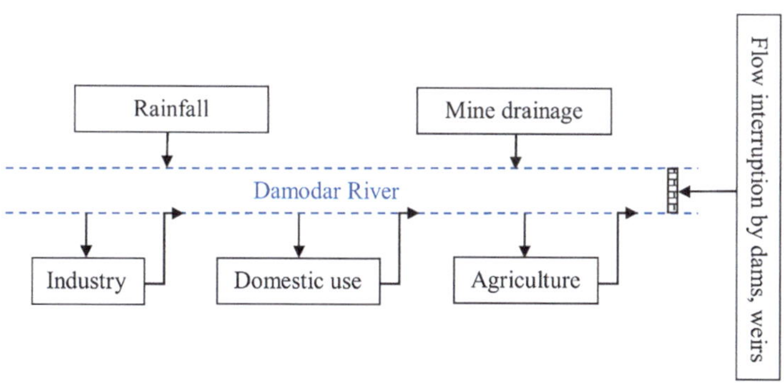

Fig. 7.1 Choking of environmental flow by anthropogenic activities based on CIFRI (2013)

7.2.2.2 Controlled River

Damodar is a typical flashy river that primarily derives its flow from immediate surface runoff resulting from rainfall, and it is notorious for its flash floods in the past. Its annual mean discharge is 12,210 million cubic meters, and about four-fifths of its flow is recorded during the monsoon period (June–September) (Patra, 2008). There is a wide seasonal fluctuation in the river flow depending on the variation of rainfall, occasional storms, intense cyclones, and resultant surface runoff. One of the perfect examples of anthropogenic intervention in the physical environment is the engineering control of the Damodar River system through the building of large dams and barrages (under DVC), expanding urban areas, and delinking the main channel from the floodplain. The total length of the Damodar is 592 km, encompassing a basin area of 23,370.98 km^2. In this span of the river valley, about 49 towns and cities, 732 live coal mines, and 131 major industrial establishments have an indirect or direct impact on modifying the physical environment of Damodar as per the estimate of the Central Inland Fisheries Research Institute (CIFRI, 1988). To minimize floods and enhance water resource management, five reservoirs and one barrage have been constructed across the rivers of the Damodar Basin to store 1270 million cubic meters of water. During 8 dry months (November to June), the river Damodar is allowed to sustain a perennial flow by regulating the reservoir sluices of DVC and Irrigation & Waterways Department (IWD). Efforts of DVC are made to maintain the flow rate of 16.14 m^3/s above Panchet Dam, and 12.17 m^3/s for the next stretch up to Durgapur and below Durgapur 2.83–5.66 m^3/s.

7.2.3 Implications for Environmental Flow Management

7.2.3.1 Environmental Flow Management with Base Flow

In environmental flow (e. flow) management, base flow refers to the minimum incessant water supply from groundwater and other delayed sources that sustains aquatic ecosystems during lean or dry periods (Gupta, 2008; Gopal, 2016). It is an important environmental flow component which affects water quality, aquatic habitat availability, and biodiversity (Gopal, 2016). Aspects associated with base flow include (1) it provides incessant habitat and water quality for aquatic life and contributes to ecosystem health through groundwater discharge, (2) maintains longitudinal and lateral connectivity for species movement and contact diverse habitats, (3) supports aquatic life by providing stable conditions, and (4) use as an indicator to assess Environmental Flow Needs (EFN), although local conditions must be considered (Chen et al., 2025; Chen & Li, 2023). It delivers the constant water needed (i.e., the minimum required quantity and quality of water) to maintain the river's health during dry seasons and periods of low rainfall (Chen et al., 2025; Hoque et al., 2022a). A continuous base flow provides essential inputs to the watershed, ensures habitat availability, supports aquatic biota, and helps maintain biodiversity and stop native destruction (Hoque et al., 2022a). Base flow sustains longitudinal (upstream-downstream) and lateral (river-floodplain) connectivity by supporting a stable water table, which is crucial for the movement of species within habitats (Hoque et al., 2022a). A continuous base water flow helps maintain water quality and retain moisture in riparian vegetation to fill the pool and avoid stagnation (Chen et al., 2025). Base flow is a vital factor of environmental flow assessment, used as an indicator for setting in-stream flow needs. Several methods, like the Eckhardt method, are used to distinguish and analyse the base flow for this evaluation (Chen et al., 2025; Chen & Li, 2023). EFNs are defined as the quantity, quality and timing of flow (i.e., environmental flow) required to maintain healthy river systems, and base flow demand or requirement varies seasonally. For example, dry-season baseflow and wet-season baseflow play different roles in natural flow systems (Chen et al., 2025). Although base flow is an important component, EF management focuses on other flow factors like peak flow, flow variability and pollution-free water to develop a comprehensive strategy for river ecosystem health (Hoque et al., 2022a; Palau & Alcázar, 2012). Hence, it needs to consider base flow requirements in hydrological planning to ensure that adequate water is supplied to support the environment, even in lean or dry periods when water demand is high (Palau & Alcázar, 2012; Gupta, 2008). It aims to ensure adequate, probable base flows to protect the ecological integrity of aquatic systems, which provide ecosystem services and support human well-being (Hoque et al., 2022a). Its assessment models and monitoring systems can be used to assess and forecast base flows, which help to set EFN standards and evaluate the effective management strategies (Chen et al., 2025; Castellanos-Osorio et al., 2025).

7.2.3.2 Environmental Flow Management on Fauna and Flora

Environmental flow management has positive effects on fauna and flora by maintaining the river flow regime with variability in magnitude, frequency, duration, timing, and rate of alteration to sustain the entire ecosystem. Significant strategies for sustainable environmental flows management are (1) determining adequate flow levels, which are important for maintaining river channels, controlling water quality and supporting fish passage, (2) implementing contamination control and habitat restoration efforts, (3) establishing ecological corridors and protected areas, (4) integrating local communities in participatory management, and (5) employing nature-based solutions. A river is a dynamic system where flow variations (high to low flow) are crucial to maintain a healthy ecosystem (Jain & Kumar, 2014). Implementation of quantitative environmental flow strategies to provide adequate flow downstream of dams and other water alterations to meet the needs of plants (flora) and animals (fauna). Implementation of both environmental flow (quantitative and qualitative) strategies to reduce water and soil contamination from urban-industrial, agricultural, and mining activities to protect habitat and wildlife populations (Hoque et al., 2022a). Releasing environmental flow can develop passages that connect fragmented habitats from downstream to upstream, and improve resilience. It can also restore impacted habitats and species damaged by human activities. Planning for environmental flows at the basin and national levels, integrating these plans for the implementation of project-level. Develop strong policies for ensuring total water use does not cross necessary limits to maintain water security and environmental health.

7.2.3.3 Environmental Flow Management in Agriculture

Environmental flow management can make water available for irrigation by increasing the water-holding capacity in existing reservoirs through the removal of siltation and the control of soil erosion. The creation of new reservoirs and wells to store water also increases water for agricultural fields. Besides, the creation of small reservoirs behind the check dams and ponds also increases water availability in the basin area, which can be used for irrigation. The check dam's installation also reduces siltation in big reservoirs, which helps to decrease siltation in these reservoirs. These creations of water-holding structures also increase local or regional groundwater through the groundwater recharge process, which can be further used for irrigation. These increasing water-holding systems or processes also raise the base flow and make water availability in river systems, especially in lean or dry periods. Besides, releasing stored water through the dams also makes water available downstream of the dam, which can be used for irrigation purposes, especially during the dry period. This water-releasing process also improves the river's water quality through the washout of pollution and increases the groundwater. The qualitative and quantitative water supply in agricultural fields is essential for crop health and crop production. These quantitative and qualitative environmental flow improvements lead to an increase in crop production as well as create potential irrigation land in the basin areas, which helps to make or promote sustainable agricultural development in the basin.

7.2.3.4 Environmental Flow Management on Groundwater

Sustainable environmental flow management can have a positive impact on groundwater through improving the quality and quantity of water in the basin area. Environmental flow management can be made sustainable for groundwater recharge through the creation of water-holding structures, such as a dam, barrage, or well and pond, and effective land use management in the basin area. Water-holding structures store water throughout the year, which influences groundwater recharge through the seepage process, leading to an increase in groundwater. These structures also lead to an increased base flow and maintain longitudinal and lateral connectivity by supporting a stable water level. Releasing water from these water-holding structures, especially from dams and barrages, leads to increased water availability downstream of the dam, particularly during dry seasons, which helps to recharge groundwater and also reduce groundwater pollution. Qualitative environmental flow management through reducing the urban-industrial, agricultural and mining effluents with heavy metals can help to decrease groundwater pollution. Moreover, reducing intensive sand mining from the river bed can promote a decrease in groundwater pollution and also increase groundwater levels. Furthermore, proper land use management planning in the basin area also helps to increase groundwater recharge and base flow. Thus, both qualitative and quantitative environmental flow management and restoration can make sustainable groundwater recharge and promote increasing the quantity and quality of groundwater, which helps to develop sustainable groundwater management in the basin.

7.2.3.5 Environmental Flow Management for SDGs

Environmental flow management is strongly associated with numerous Sustainable Development Goals (SDGs), especially SDG 6 (clean water and sanitation), which is needed to protect and restore aquatic ecosystems. It supports other environmental goals like SDG 14 (life below water) and SDG 15 (life on land) by sustaining the health of freshwater and estuarine ecosystems. It also supports SDG 1 (no poverty), SDG 2 (zero hunger) and SDG 8 (decent work and economic growth) by ensuring sustainable fisheries and livelihoods. Besides, it supports SDG 3 (good health and well-being) by maintaining water quality and ecosystems, which contribute to human well-being. It also helps to achieve SDG 12 (responsible consumption and production). Environmental flows are crucial to protect and restore aquatic ecosystems such as rivers, estuaries and wetlands, to reach SDG 6 by ensuring the health of these ecosystems. Excessive withdrawal leads to water stress, which measures against available renewable water resources, as EFs help determine the amount of water required for ecosystem sustainability (Dickens et al., 2025). Besides, EF maintains freshwater and estuarine productivity of fish, molluscs and other aquatic organisms, which are important components of aquatic ecosystems, supporting underwater life (SDG 14) (Ramsar Policy Brief on Convention on Wetlands, 2018). It supports healthy aquatic ecosystems, which contribute to overall terrestrial biodiversity and ecosystem integrity, supporting life on land (SDG 15) (Yang et al., 2020). Moreover, EF provides sustainable fisheries and other aquatic species, which ensure food security and economic activities for local communities, contributing to the achievement of SDG 1 (no poverty) and SDG 2 (zero hunger) (Ramsar Policy Brief on Convention on Wetlands, 2018; Yang et al., 2020). It helps sustainable resource management through sustaining and maintaining ecological flows and supports long-term economic activities which depend on water, including fisheries and tourism, leading to the achievement of SDG 8 (decent work and economic growth) (Mallick et al., 2025). Furthermore, it also contributes to the overall welfare of the people through maintaining water quality and ecosystem health. Its management is important to ensure water-use proficiency and sustainable extraction through integrated water resources management (IWRM), helping to support SDG 12 (responsible consumption and production). Hence, environmental flow management and restoration will help achieve several SDGs, including SDG 6.

7.3 Integrated Environmental Flow Restoration

Environmental flow restoration can be achieved through quantitative and qualitative techniques. Quantitative techniques help achieve the minimum quantity of water flow while qualitative techniques are concerned with quality water availability (Fig. 7.2). These strategies are based on robust datasets and methodology as discussed in the following sections. Primary level field survey at the selected villages and secondary data mainly from DVC annual reports (2020–2023), and the district census handbook (2011). The local people's perception regarding environmental flow restoration of the river has been taken by a direct field investigation method.

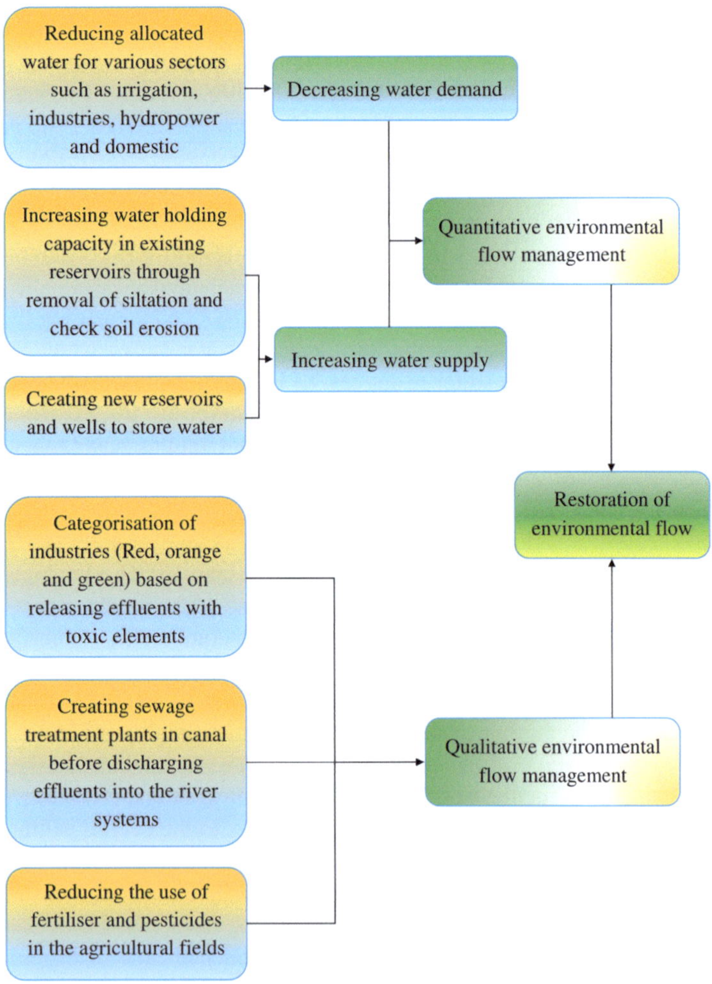

Fig. 7.2 Schematic diagram for flow restoration strategy

Besides, the physical measurements of the check dam and dug well are also done by instrumental survey.

The identification of discharging effluent *nallahs*, location of water intake point, dug well, sewage treatment plant (STP), industry, and mining area, and the measurement of the sand mining area has been done through Google Earth Pro (2023–24). Moreover, land use and land cover maps of the selected two villages (Urwan and Rajarampur) for environmental flow restoration management strategy were also prepared using Google Earth Pro (2024). Statistical techniques, like mean, median, and range, are also used in this study.

7.3.1 Quantitative Restoration Strategies

Quantitative e. flow management aims at increasing water supply throughout the year through the augmentation of more water into the channels, especially in the lean months (December–May) of the year. The present study has indicated three alternatives for quantitative flow management as discussed in the following sections.

7.3.1.1 Increasing the Dam-Reservoir Capacity Building

Dam reservoirs exhibit a lowering of the storage capacity due to the gradual siltation of the reservoir over time. For example, about 55% dead storage capacity and 33% live storage capacity of the Maithon and Panchet dams were reduced till 2020. In this context, two alternatives were conceived—(1) the construction of new reservoirs for storing monsoon water and (2) the maintenance of the existing reservoir. Regarding the first alternative, the Balpahari reservoirwas proposed at Giridih, Jharkhand, but it has not seen the light since its proposal in 2014. One of the major hindrances to its implementation is the massive alteration of the ecology, the issue of fertile agricultural land submergence, rehabilitation, and huge monetary involvement. Thus, the construction of a new reservoir has been debated and not executed to date. The second alternative, i.e., desilting through dredging to increase the capacity of the reservoir to store more water during the monsoon period is also critical. For this purpose, DVC assessed that about INR 50,000 crore is required for the total desiltation of both dams (Maithon and Panchet) (Banerjee, 2021). DVC also mentioned that the problem of desiltation is not techno-economically viable, but the problem is related to the management of excavated earth disposal and the higher cost related to repeating the desiltation process. The dead storage of Maithon and Panchet is 90 and 107 MCM respectively. Therefore, among the 197 MCM dead storage, 108.35 MCM has been reduced.

It has been shown that about INR 9090 crore is required for desilting 10% of total dead storage while INR 6000 crore is required for constructing the proposed Balpahari Dam at Giridhi (Jharkhand) which is proposed to be constructed 60 km upstream of the Maithon Dam and also covers about 86% of the catchment area of the Maithon dam. This project escalates the life of the Maithon Dam, but the construction is not initiated due to issues concerning the administrative, financial, and technical aspects. It appears that cost-benefit analysis will neither support new reservoir construction nor dredging of the existing reservoir. It is observed that the costing of desiltation to increase by 20% the dead storage capacity of Maithon and Panchaet requires INR 18,181 crore, and to increase by 50% capacity requires about INR 45,454 crore.

The calculation shows that the monetary involvement for desiltation is quite high, even higher than in constructing the new reservoir (e.g., the cost of the Balpahari dam proposed reservoir). Therefore, small-scale management requires less monetary involvement engaging the local community may reduce the siltation rate by checking the soil erosion through community watershed management. For example, the plantation of native species like *Sal, Simul, Palash, and Mahua* is required to check soil erosion instead of eucalyptus, which induces more soil erosion in the basin (Amenu, 2017). Plantation could be done in a social forestry manner using the local community that will provide food, fodder, and fibre besides soil erosion check and environmental degradation. The field survey also indicates the interest of the local people in engaging themselves in this project. The DVC also initiated some measures like the construction of structures (check dam, silt detention, farm pond and gully control and drainage line), afforestation, and upland treatment. About 214,963 ha area has been afforested of which 42,149 ha area on the farmer wasteland. Upland treatment was taken up in 305,400 ha of land and 71,969 ha of wasteland has been reclaimed.

7.3.1.2 Check Dam Management

The small check dams in the study area are escalating with time (Fig. 7.3a). These check dams are mainly constructed by the DVC to check soil erosion in the monsoon period. At present, there are more than 17,000 check dams of which most of the check dams are dry (DVC, 2022). As there is a small reservoir behind the check dams, that reservoir may also store some water for agricultural use (Fig. 7.3b, c). These small structures built up a water storage capacity of 25,701 ha-m which has the potential to irrigate 67,225 ha.

However, many of the reservoirs with check dams are silted up denying storing of excess monsoon water in the check dams. The DVC also initiated reviving the old structures and to date, 1624 old structures have been restored. Similarly, desiltation work is in progress for the 1912 check dams for their new life. The present study also examined several check dams that suffer from maintenance from the perspective of water storage. Despite the repairing process initiated by the DVC, the local community can also be involved and at the community level, the repairing process can be resolved immediately. During the field survey (2023–24), the local people also opined that a 100-day Mahatma Gandhi National Rural Employment Guarantee

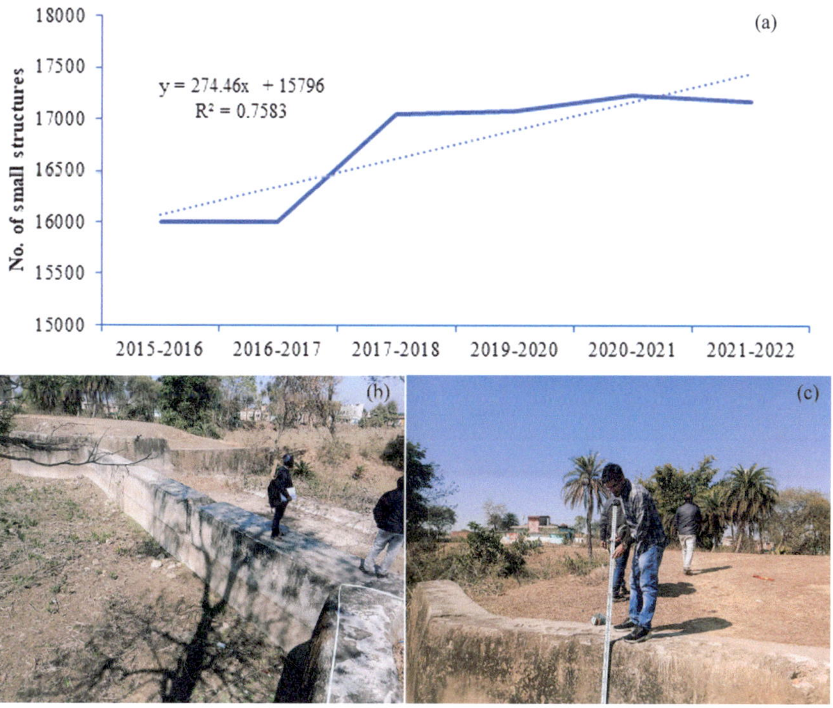

Fig. 7.3 Small structures in DRB; (**a**) The growth of the small structures (check dams) in the DRB from 2015–16 to 2021–22 (Note: data for the year 2018–2019 is not available), (**b**) Check dam situated in the Urwan village with a length of 22.67 m and width of 2.62 m, (**c**) Measurement of check dam by the authors

Act (MGNREGA) project fund could be utilized for check dam desilting at the local level.

Considering the limitations of the massive reservoir, DVC has also increased the number of check dams. The check dams are mainly constructed in the upper basin area. As per the estimation of Dashora et al. (2019), the cost of each check dam constructed from 1995 to 2005 varied from INR 150,000 to 500,000. In the year 2022, the cost of one check dam construction is 10 lakh to 1.5 crore based on the size of the check dam (The Times of India, 2022). Apart from the construction, the maintenance charge regarding the dredging of silt and repair costs is also to be added. Therefore, it is a proposal to civil engineers to think of the alternatives of check dam. Thus, what is more important is to maintain the check dams instead of mushrooming the check dams. Moreover, small watershed management like changes in the cropping pattern like the introduction of cover crops instead of open crops, and minimal disturbance of the soil, especially in the riverine buffer (Zafirah et al., 2017). The villagers also agreed to small-scale interventions rather than large-scale projects.

7.3.1.3 Local People's Strategies for Water Resources Management

Local and indigenous knowledge and practice are important for sustainable water resource management at the village level. The field survey in the study villages located in the DRB portrays two different types of practices. The first is the community water resource management at the upper DRB dominated by the hard rock terrain. The second is related to the community water resource management at the lower DRB dominated by alluvial terrain as discussed in the following sections.

(a) Community Water Resource Management at the Upper Damodar River Basin

The field survey at the upstream villages like Urwan in the Koderma district exhibits some fascinating observations. Local people said that they do not benefit from DVC because they don't get water for irrigation, or electricity for their consumption. When they were asked about the river e. flow and river restoration, they clearly indicated that the river must be saved for their needs. If the river is of no use in terms of irrigation, or electricity facility, they are not bothered about river restoration. However, another group of villagers opined that if the river does not exist, no future development is possible. Thus, saving the river is necessary. Remote sensing, aided by field surveys, indicates that huge river water is used for different purposes e.g. domestic, agricultural, and small industry. Out of the 132 water intake points, 95 were concentrated in the upper part of the DRB (the Barakar and main Damodar Rivers) for local water use (Fig. 7.4). During the lean period the local industry, urban areas also use the reservoir water through the water carrying tankers (Fig. 7.5d).

The reservoir is constructed at the expense of the agricultural land, they vehemently expressed their strong desire to get irrigation water through pipe from the reservoir. The local people excavated the small drainage line from the reservoir to the nearby agricultural land and the water is then used in the farmland through the pipeline (Fig. 7.5a, b). At present, there are three existing systems of household and irrigation water-(1) dug well and (2) *chapa kol* (tube well), (3) water supply from a water plant at the cost of INR 60 per month. Dug wells are constructed at their own cost while the *chapa kol* is installed by the local panchayat (Fig. 7.5f). They said that digging a well in this hard rock terrain costs around INR 4 lakh. The depth of the well is 15 m with a 1.75 m radius and 1.25 m thickness of wall. The well water is coming from the seepage from the reservoirs, and surrounding areas. However, wells get dried up during the lean months (April–May). Local people urge for the maintenance of the wells for getting water for household and irrigation purposes. Moreover, they also demanded that such types of wells could be constructed for every household by the government and other industrial and business firms. Observing the locality, it is also felt that there is a scope for constructing dug wells in the study area. The total number of households in the Urwan village was 776 as per the District Census Handbook (2011). Out of the 130 wells, 88 wells are found

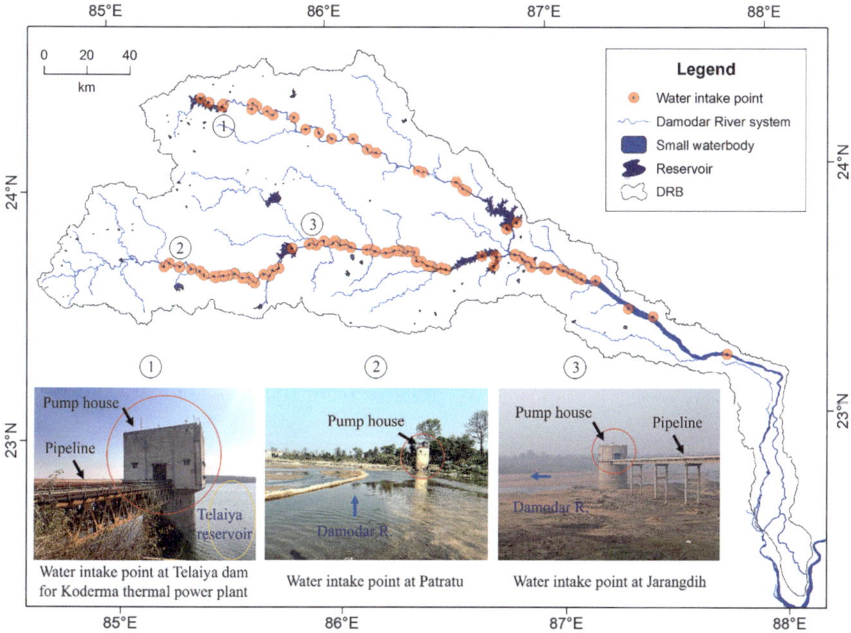

Fig. 7.4 The location of water intake points in main Damodar and Barakar River and field photos of pump houses taken in February 2024

on agricultural land and 42 are mainly used for domestic purposes (Figs. 7.5e and 7.6). Therefore, as per the demand, another 734 (776–42) wells (taking 2011 census households which may increase while diagnostic survey and execution of the plans) can be installed and the probable cost would be around INR 2936 lakh. The local people agree to invest about INR 0.5–1 lakh per well and the rest can be borne by the government and industrial and business firms. The local people also indicated that the industrial and business firms extracting the water from the reservoir without any tax for selling the water for urban, and industrial purposes, might share a certain well construction cost (INR 0.2–0.5 lakh per well). In other words, villagers want INR 1 lakh will be invested by the local people (beneficiary), industrial, and business firms. The rest (~3 lakh per well) would be shared by the government. If the government takes this initiative to construct one dug well per household, the stress upon the river water can be minimized because household wells are also used for irrigation. It is seen that the excess water could store in the well during the monsoon period. Some parts will recharge groundwater and the rest water will be available for household consumption throughout the year except for two lean months (April–May). However, plan layout, material composition and technological innovations may also increase the capacity of wells for assuring the year-round household water supply.

Another similar type of structure is the construction of the tanks at the community level which will be helpful for irrigation and household water requirement.

Fig. 7.5 Community water resource management; (**a**) and (**b**) Denote excavation of canal-like drainage by local people for irrigational purposes in Urwan village, (**c**) Reservoir adjacent agricultural land, (**d**) Water extraction from the reservoir for industrial use in the lean season, (**e**) Water extraction from dug well and (**f**) *Chapa kol* in Urwan village

Tanks were designed and constructed 2000 years ago by local rulers and maintained by the local communities in southern India (Chinnasamy & Srivastava, 2021). Tank irrigation is mainly observed in the arid and semi-arid regions of the southern part of India mainly characterized by undulating topography. This study region mainly the upper part is vulnerable to water scarcity as the region is solely dependent on rainfall. Therefore, the tank—the water-holding structure may also help achieve the sustainable development goal (SDG 6), i.e., concerning overcoming the challenges of achieving surface and groundwater security and nourishing adequate water

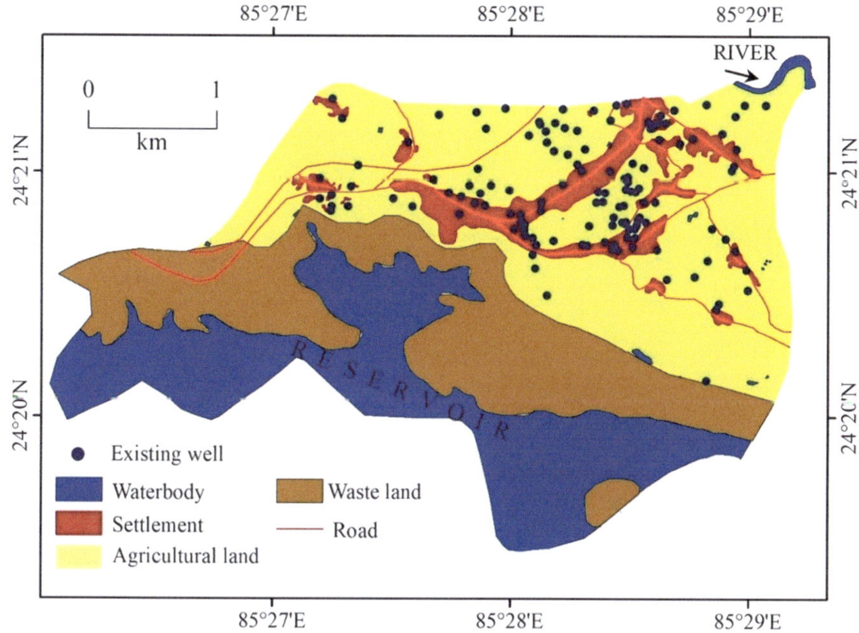

Fig. 7.6 Location of existing dug well in the Urwan village

supply (Benson et al., 2020). This structure is mainly used for rainwater harvesting. Local people of the present study area are also keen to get this type of tank from government initiatives. They mentioned that land is available at the hands of DVC that could be used for this purpose. The tank was already constructed in some places in the Urwan village mainly for irrigation purposes. The DVC also renovated rainwater harvesting structures like tanks (surface water storage at the hard rock region, i.e., upper DRB) and ponds (surface water storage at the alluvial region, i.e., lower DRB) specified in Table 7.3.

The water demand for the study villages has been assessed with the help of field data (Table 7.4). Apart from domestic use, the requirement for irrigational water is also stated in Table 7.5. Therefore, the community tanks could potentially be used for both household water supply and irrigation. For this type of rainwater harvesting, there are a few challenges and scope. For example, drainage of clean water to tanks is necessary. In a study village like Urwan, there is rarely a toilet found; mostly they are accustomed to open defecation that is to be stopped completely. Otherwise, water from the tank will not be useful for meeting household water demand. Central government interventions are already there through *Swachh Bharat Mission* to stop open defecation but still, rural people do it due to the lack of consciousness and traditional belief of not having a toilet connected to the living room. Even they are reluctant to construct toilets in the vicinity of the households. Construction cost is another challenge.

Table 7.3 Renovation and construction of tanks and ponds by DVC

Year	Renovation of pond			Construction of new rainwater harvesting structures		
	Number of ponds renovated	No. of people benefitted	Hectare of land cultivated	No. of new rainwater harvesting structures	No. of people benefitted	Hectare of land cultivated
2019–2020	120	5000	1200	35	400	350
2020–2021	120	4500	1000	38	500	400
2021–2022	123	3100	1150	37	400	350
2022–2023	120	2000	2770	35	400	350

Source: DVC (2020, 2021, 2022, 2023)

Table 7.4 Demand for water for domestic purposes (L/cap/day)

Village	Total households	Drinking	Cooking	Bathing	Cloth washing	Domestic animal	Toilet and sanitation	Total water use
Urwan	776	2.96	6.65	18.25	9.65	3.15	11.2	51.86
Urro	232	3.01	8.84	17.58	7.58	2.15	12.5	51.66
Landhadi	51	2.84	7.29	12.54	8.75	3.09	11.25	45.76
Lalpania	723	3.25	5.68	13.52	9.85	1.67	24.85	58.82
Khetko	1510	2.98	7.89	18.59	10.58	1.28	10.85	52.17
Lohpiti	634	3.02	8.81	14.57	7.85	2.58	8.89	45.72
Sheopur	163	2.54	10.25	14.53	6.14	3.05	15.64	52.15
Hijuli	1498	3.26	8.96	24.25	7.45	1.25	26.08	71.25
Arddhagram	254	3.02	10.5	15.85	6.58	2.34	12.25	50.54
Pratappur	161	2.89	7.95	10.25	4.85	3.25	13.57	42.76
Randiha	459	2.99	9.25	18.95	5.78	1.12	13.25	51.34
Rajarampur	444	2.86	7.89	17.84	8.75	2.5	8.82	48.66

Computed by the authors based on field survey (2023–2024)

Table 7.5 Irrigation requirements for crops in different months

Month	Irrigation requirements of the crops (in inches) on the basis suggested by the Ministry of Food & Agriculture	Irrigation requirement at canal head (inches)	Total requirement at canal head 103 acre ft
June	0.2	0.286	24
July	4	5.714	467
August	2.2	3.143	255
September	4	5.714	467
October	5	7.143	579
Total	15.4	22	1792

Source: Sinha and Rao (1985)

The construction cost of the tank with a capacity of 5000 m^3 is nearly INR 2 lakh in the hard rock terrain in the year 2007 (Mishra et al., 2007). Thus, local people demand that the government would construct these tanks on the DVC land. They are also ready to pay a certain amount for this purpose. Another important dimension is related to integrating the tanks with the dug well. A study conducted by Mishra et al. (2007), depicts that tank-dug well integration is possible if they are constructed in a valley with a lower slope (2–5%) having a pattern-denoted by tanks at the upper part of the slope followed by the dug wells. The seepage water from the tanks may recharge the dug wells located below the tanks. As per their findings, even if tanks dried up, wells would contain water in the lean months. However, the site suitability and cost-benefit analysis are vital at the ground level before finalizing the draft plans and their execution.

(b) Community Water Resource Management at the Lower DRB

The water resource management at the community level in the alluvial terrain is different from the hard rock-dominated upper DRB areas. To examine the typical traits of this community water management, a case study has been done in the Rajarampur village located as the southernmost study village. About 90% of the area is acquired by agricultural land, the agricultural land located at the edge of the river uses the river water while most of the farmers use groundwater for agricultural purposes (Fig. 7.7). Local people are already using the ponds for the domestic

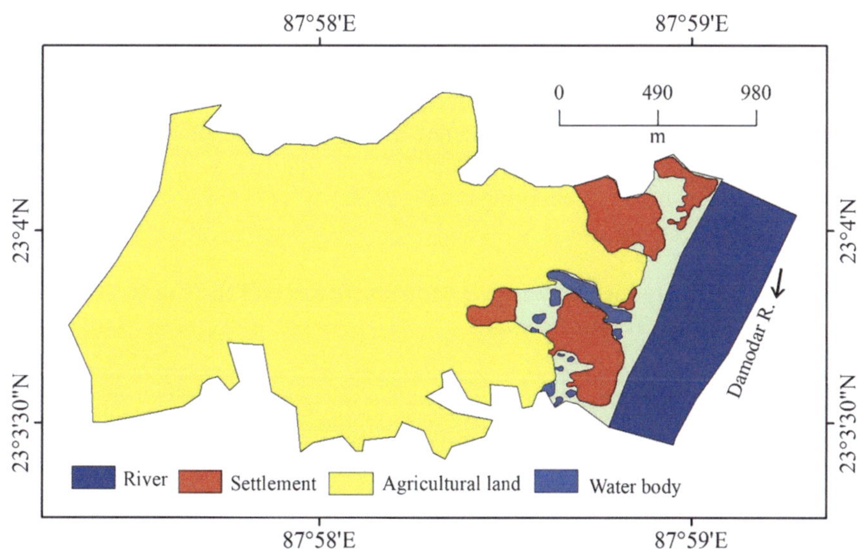

Fig. 7.7 Land use and land cover of Rajarampur village in Jamalpur CD block, *Purba* Bardhaman district sows that huge expansion of agricultural land with almost no ponds. This indicates the potentiality of pond excavation in the study village

purposes which are locally known as 'doba' '(small pond with an area ~100 m^2) where the local people capture the rainwater or store the excess water from the DVC canal. Another type of large community pond known as *'sech pukur'* (~100,000 m^2) is mainly used for agricultural purposes (Mahammad, 2022). The local community is aware of the diverting of surplus water to the pond through the canal during monsoon months (July to September). Field surveys clearly indicate their eagerness to use the excess floodwater in the ponds. The rural people also proposed the construction of more *'sech pukur'* with the help of local administration as the land is available in the locality. This type of community pond may be used for two main purposes—(1) storing the excess floodwater and using it during the lean period for agricultural purposes, and (2) recharging the groundwater to restore the groundwater level. The local community is also ready to pay the initial cost for constructing the *'sech pukur'*. About 75% of respondents are willing to pay INR 6000 at one time and they are eager to the govt for the land acquiring process for the generation of community ponds. The local people also express their eagerness for the use of excess floodwater passing through the DVC canal and also enrich the *doba* and *sech pukur*. After the interaction with local people, they are also eager to reduce the *boro* cultivation and cultivate the less water-intensive crops.

7.3.2 Qualitative Restoration Strategies

The previous section is dedicated to quantitative e. flow management while this section is concerned with water quality improvement. The Damodar River has deteriorated because of different issues, i.e., discharging effluents from *nallahs*, irregular monitoring of industrial wastewater, and intensive mining activities. This section indicates the probable pathways for qualitative restoration of e. flow. Regarding these three pathways have been considered, i.e., (i) identifying discharging effluent *nallahs* and construction of STP, (ii) monitoring of industries with an emphasis on green initiatives, and (iii) monitoring mining areas.

7.3.2.1 Identifying the Discharging Effluent *Nallah* and the Construction of Sewage Treatment Plant

The industrial discharge and municipal and agricultural effluents are drained through the *nallahs* of the basin area and such *nallahs* are meeting with the river and deteriorating the water quality. Therefore, the identification of such effluent *nallahs* along with its treatment in a basin is important for improving the water quality of the river. In the Damodar River Basin, a total of 21 discharging effluent *nallah* have been identified (Fig. 7.8).

Different processes are applied for the treatment of the *nallah* like bar screens, grit chamber, primary sedimentation tank, trickling filter, etc. The effluents

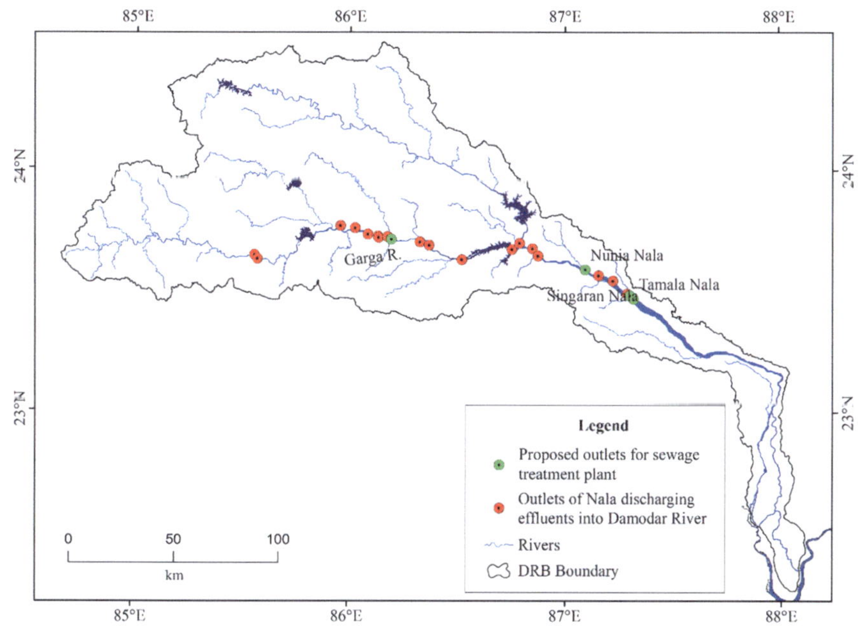

Fig. 7.8 Distribution of outlets of discharging effluent *nallahs* and proposed locations of sewage treatment plant in the Damodar River Basin

containing *nallah* of DRB pass the urban-industrial effluents. The Asansol-Durgapur urban-industrial effluents are discharging through Tamala, Singaran, and Nunia *nallah* into the river. The extensive field survey coupled with observations from different reports, four *nallahs* are identified as the vital *nallah,* i.e., Tamala, Singaran, Nunia, and Garga. Therefore, the installation of the STP has been proposed in these four outlets. Now the sewage installation cost is the prime concern. The cost of construction of STP depends upon the capacity and it ranges between INR 4–80 lakh or in crore depending upon the sewage generation. The Nasik municipal commission estimated the cost of STP along with the weir and pumping station is 130.11 crore for a capacity of 130 MLD (1 million litres = 1 MLD) (Anparthi, 2013). The cost of installation of STP may be shared by the DVC and state government and central government, local people.

7.3.2.2 Monitoring of Industries with an Emphasis on Green Initiatives

The DVC area is famous for its coal-based thermal power plant, hydropower plant, and numerous small-scale industries. The industrial belts are responsible for polluting the environment. For the purpose of monitoring the industries, the Central Pollution Control Board (CPCB) classifies the industrial sectors as red, orange, green, and white based on the magnitude of pollution. The red category signifies the

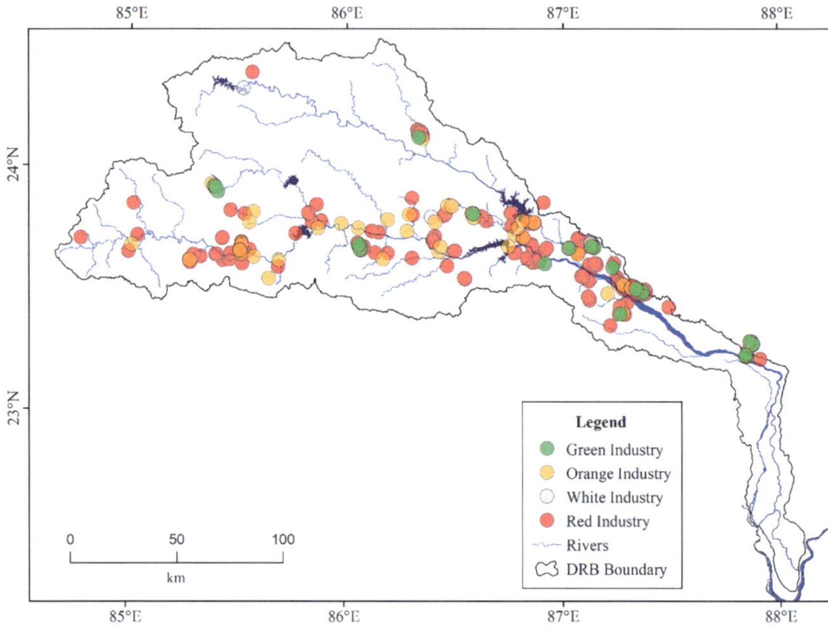

Fig. 7.9 Distribution of different types of industry in terms of pollution

most polluting industry having a pollution index of 60 and above, 41–59 is orange, 21–40 is green, and equal or up to 20 is white industry (CPCB, 2016).

The DRB exhibits a total of 292 industries out of which 27, 53, 3, and 209 are green, orange, white, and red respectively (Fig. 7.9). It has been observed that the red industry dominates the basin, therefore regular monitoring of the red industry is needed. Initiatives taken by the industry and government regarding periodical monitoring and publishing the status of the pollution should be prioritized. The polluted water from industry meeting with the river through different effluent *nallahs* deteriorated the river's health to a large extent and affected the aquatic community. To restore the qualitative environmental flow, further industrial development should prioritize green industries. Similarly, further development of the red and orange industry in the basin area should be minimized. Moreover, the rules and regulations of the CPCB especially for regular monitoring of the red industry regarding wastewater management shall be strictly followed for minimizing the pollution level.

7.3.2.3 Monitoring Mining Areas

Mineral mining and sand mining activities are both escalating in the DRB. Open cast and underground mining are observed in this basin (Fig. 7.10). Two underground mining areas have been found in the basin area, i.e., Ray Bachra Colliery U/G and Eastern Coalfields Limited (ECL) river bed mining. About 46% coal

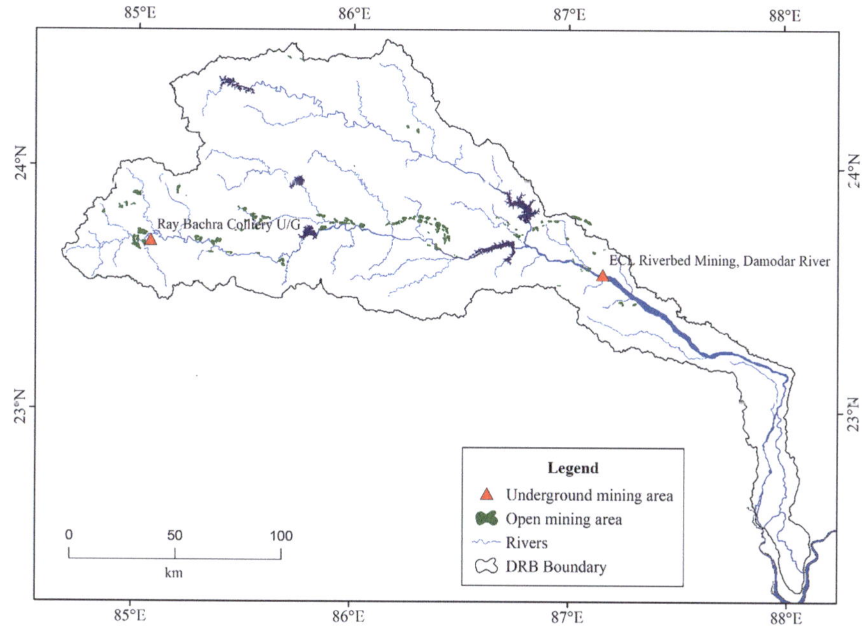

Fig. 7.10 Distribution of underground and open mining area

reservoir of India is situated in the DRB (Tiwary & Dhar, 1994). Major coal fields in the basin are located in Raniganj, Jharia, and Bokaro. Mining water and coal washery effluents alter the chemical composition of the river water (Tiwary & Dhar, 1994). Mining water exhibits higher levels of SO_4^{2-}, hardness, and bacterial contamination while coal washery water contains a higher TSS, iron, oil, and grease. The increasing TSS in the form of fine coal particles increases the TDS and COD of Damodar River water (Tiwary & Dhar, 1994). Therefore, well-designed disposal sites, installation of wastewater treatment machines, and recycling and reuse of treated water may help to increase the quality of e. flow from both quantitative and qualitative perspectives.

Sand mining with heavy machines and vehicles is also a major problem for the Damodar River. Ghosh et al. (2022) mentioned that intensive in-channel sand mining was one of the factors leading to poor fluvial functionality of the Damodar River upstream and downstream of the Durgapur Barrage. Figure 7.11a shows that almost 80% of river stretches of the Damodar and Barakar Rivers are mined. The intensity of mining is higher in the middle Damodar (Fig. 7.11b). Bhattacharya et al. (2019) found that turbidity, TDS, and DO are higher in the mining area exceeding the standard value of BIS and WHO. They also observed that instream sand mining negatively impacts the aquatic community and decreases the species richness and diversity.

The three pathways considered in this section are not separated from each other but rather integrated in a participatory manner. Therefore, all the stakeholders

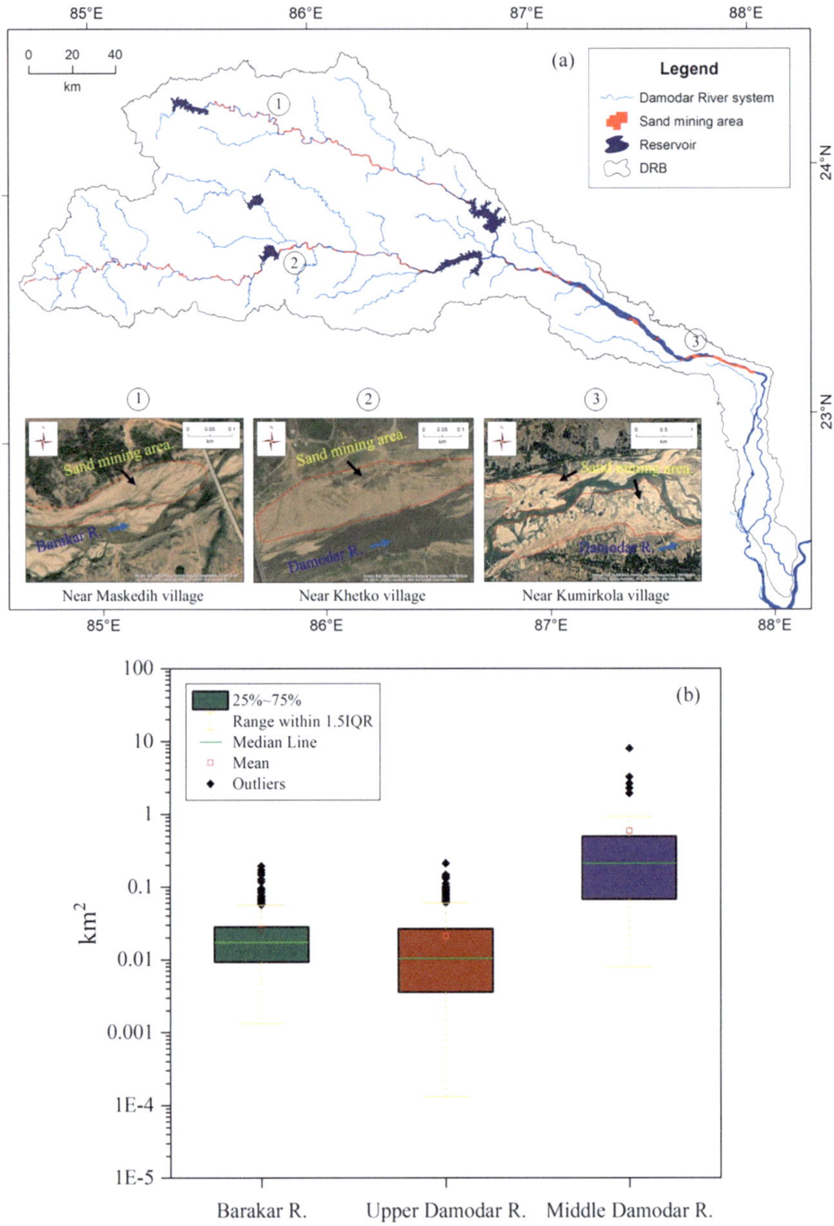

Fig. 7.11 Intensity of sand mining; (**a**) Location of sand mining stretches of the main Damodar and Barakar River and (**b**) Amount of sand mining in the Damodar and Barakar River

(industry, state and central government, and local community) may adopt the appropriate initiatives for developing an integrated participatory framework for the sustainability of the Damodar River Basin.

7.4 Concluding Notes

Implementation of environmental flows in river systems is a challenging matter due to various reasons. The first problem is related to flow availability in the reservoir. It needs the volume of water, the right frequency, the right time, and the right length of flow time to sustain a healthy river and maintain the ecological health of river systems. However, in the context of DRB, the volume of water in the reservoirs is gradually dwindling with time due to siltation in the reservoirs, as mentioned earlier. Second is the Cumulative Impact Assessments (CIAs) of dams, certification, and maintenance of law. Certification of industrial development infrastructure and land use, water use, and energy production strategies is essential. Environmental impact assessment on all developmental projects, such as hydropower generation, water extraction for irrigation, industrial, domestic, and other uses, including water regulation, should be made mandatory for estimating environmental flows. These development projects must obey environmental protection laws and maintain minimum environmental flow in aquatic systems, especially during the dry period. However, in the context of DVC and CIA, certification and obedience to laws are rare. Field visits show that illegal water extractions from the reservoirs/water intake points are clearly visible in the study villages, like Urwan in Giridh, Jharkhand.

Besides, integrated planning with proper execution greatly influences the environmental flow restoration. Better planning, design, and operation of the dam should minimize negative impacts and indicate where to build or not to build it. The proposal for the construction of new reservoirs needs critical introspection using a multi-criteria assessment. As DVC is an interstate river basin, integrated planning incorporates the state govt. of West Bengal, Jharkhand, and the central govt. (DVC) is needed. However, due to the differences in the state policies, the uniformity in the planning and execution is not reached in many cases, which hinders the integrated watershed management practices. Its acceptance and implementation are incorporated into water policies and laws of the IUCN, it needs to be integrated into sustainable water management by making the liability of the government, ministries, and other stakeholders. Achieving sustainable development downstream requires rising care by countries, especially the developing nations, both by public and private sector investments.

References

Amenu, B. T. (2017). Review on impact of eucalyptus plantation on the soil. *International Journal of Scientific Research in Civil Engineering, 2*, 37–43.

Anparthi, A. (2013). STP's project cost vary in different schemes. *The Times of India*. Retrieved September 8, 2022, from https://timesofindia.indiatimes.com/city/nagpur/stps-project-cost-vary-in-different-schemes/articleshow/21063105.cms

Banerjee, S. (2021). Damodar basin water management system—A review on critical issues and performance. In *ICOLD Symposium on Sustainable Development of Dams and River Basins, New Delhi*.

Benson, D., Gain, A. K., & Giupponi, C. (2020). Moving beyond water centricity? Conceptualizing integrated water resources management for implementing sustainable development goals. *Sustainability Science, 15*(2), 671–681. https://doi.org/10.1007/s11625-019-00733-5

Bhattacharya, R. K., Das Chatterjee, N., & Dolui, G. (2019). Consequences of sand mining on water quality and instream biota in alluvial stream: A case-specific study in South Bengal River, India. *Sustainable Water Resources Management, 5*(4), 1815–1832. https://doi.org/10.1007/s40899-019-00345-y

Castellanos-Osorio, G., Abbas, S. A., Molina-Navarro, E., Bailey, R. T., & Senent-Aparicio, J. (2025). Assessing improvements in environmental flow estimation through the use of a coupled surface-groundwater model. *Groundwater for Sustainable Development, 29*, 101426. https://doi.org/10.1016/j.gsd.2025.101426

Chen, H., & Li, Q. (2023). Testing and applying baseflow approaches to environmental flow needs. *Ecological Indicators, 152*, 110363. https://doi.org/10.1016/j.ecolind.2023.110363

Chen, H., Li, Q., Wang, Q., Chai, X., & Wang, Y. (2025). Applying baseflow approach to the environmental flow needs of the Similkameen River Watershed in British Columbia, Canada. *Journal of Hydrology: Regional Studies, 57*, 102137.

Chinnasamy, P., & Srivastava, A. (2021). Revival of traditional cascade tanks for achieving climate resilience in drylands of South India. *Frontiers in Water, 3*, 639637.

CIFRI. (1988). *The river Damodar and its environment*. Indian Council of Agricultural Research, Barrackpore, Bull. No. 73. http://www.cifri.res.in/Bulletins/Bulletin%20No.79.pdf

CIFRI. (2013). *Environment and fisheries of river Damodar*. Indian Council of Agricultural Research, Barrackpore, Bull. No. 183. http://www.cifri.res.in/Bulletins/Bulletin%20No.183.pdf

CPCB. (2016). *Ministry of Environment and Forest*. Government of India. Central Pollution Controll Board (CPCB). https://cpcb.nic.in/uploads/Latest_Final_Directions.pdf

Dashora, Y., Dillon, P., Maheshwari, B. L., Soni, P., Mittal, H., Dashora, R., et al. (2019). Hydrologic and cost benefit analysis at local scale of streambed recharge structures in Rajasthan (India) and their value for securing irrigation water supplies. *Hydrogeology Journal, 27*, 1889–1909.

Dickens, C., Tharme, R., Biancalani, R., Marinelli, M., & McCartney, M. (2025). *Incorporating environmental flows into "water stress" indicator 6.4. 2*. Food and Agriculture Organization of the United Nations. Retrieved September 25, 2025, from https://www.unwater.org/sites/default/files/app/uploads/2019/01/SDG6_EF_LOW2.pdf

District Census Handbook, Directorate of Census Operation. (2011). India. Retrieved March 15, 2019.

Dutta, V., Sharma, U., & Kumar, R. (2020). Restoring environmental flows for managing river ecosystems: Global scenario with special reference to India. In *Environmental concerns and sustainable development: Vol. 1: Air, water and energy resources* (pp. 163–183).

DVC. (2020). *Annual Report 2019–2020*. Retrieved August 9, 2021, from https://www.dvc.gov.in/cms-web/annual_report

DVC. (2021). *Annual Report 2020–2021*. Retrieved August 9, 2021, from https://www.dvc.gov.in/cms-web/annual_report

DVC. (2022). *Annual Report 2021–22*. Retrieved December 18, 2023, from https://www.dvc.gov.in/cms-web/annual_report

DVC. (2023). *Annual Report 2022–23*. Retrieved January 4, 2024, from https://www.dvc.gov.in/cms-web/annual_report

Ghosh, S., Hoque, M. M., Saha, U. D., & Islam, A. (2022). Assessment of dam-induced changes in ecogeomorphological behaviour and fluvial functionality in the Damodar River, West Bengal, India. *AQUA—Water Infrastructure, Ecosystems and Society, 71*(6), 722–750. https://doi.org/10.2166/aqua.2022.003

Gopal, B. (2016). A conceptual framework for environmental flows assessment based on ecosystem services and their economic valuation. *Ecosystem Services, 21*, 53–58. https://doi.org/10.1016/j.ecoser.2016.07.013

Gupta, A. D. (2008). Implication of environmental flows in river basin management. *Physics and Chemistry of the Earth, Parts A/B/C, 33*(5), 298–303. https://doi.org/10.1016/j.pce.2008.02.004
Hoque, M. M., Islam, A., & Ghosh, S. (2022a). Environmental flow in the context of dams and development with special reference to the Damodar Valley Project, India: A review. *Sustainable Water Resources Management, 8*(3), 62. https://doi.org/10.1007/s40899-022-00646-9
Hoque, M. M., Islam, A., & Mahammad, S. (2022b). Assessing the surface and bottom river water quality for drinking purpose and human health risk analysis: a study of Damodar River, India. *Arabian Journal of Geosciences, 15*(23), 1734. https://doi.org/10.1007/s12517-022-10987-6
Hoque, M. M., & Islam, A. (2024). Spatio-temporal dynamics of ecological, bacteriological, and overall water quality of the Damodar River, India. *Environmental Science and Pollution Research, 31*(12), 18465–18484. https://doi.org/10.1007/s11356-024-32185-5
Hoque, M. M., Islam, A., Islam, A. R. M. T., Pal, S. C., Mahammad, S., & Alam, E. (2023). Assessment of soil heavy metal pollution and associated ecological risk of agriculture dominated mid-channel bars in a subtropical river basin. *Scientific Reports, 13*(1), 11104. https://doi.org/10.1038/s41598-023-38058-0
Hoque, M. M., Islam, A., Sarkar, B., & Saha, U. D. (2024). Assessing the surface and bottom river water quality for irrigation: A study of Damodar River, India. *International Journal of Energy and Water Resources, 8*(4), 545–562. https://doi.org/10.1007/s42108-022-00206-z
Hoque, M. M., Islam, A., Pal, S. C., Sarkar, B., & Hossain, Z. M. (2025a). Assessment of fish habitat suitability in the wake of elevated pollution of a tropical river in India. *Environmental Science and Pollution Research, 32*(3), 1208–1222. https://doi.org/10.1007/s11356-024-35685-6
Hoque, M. M., Pal, S. C., Mandal, S., & Islam, A. (2025b). Assessment of mollusc habitat suitability of an anthropogenically altered tropical river in India. *Environmental Monitoring and Assessment, 197*(10), 1–18. https://doi.org/10.1007/s10661-025-14593-3
Islam, A., Sardar, N., Mohinuddin, S., Hoque, M. M., Sengupta, S., Das, B. C., ... & Sengupta, B. (2023). Quasi-equilibrium channel metamorphosis in planform of a subtropical river in India in post-dam period. *Catena, 221*, 106793. https://doi.org/10.1016/j.catena.2022.106793
Islam, A., Das, B. C., Mahammad, S., Hoque, M. M., Pham, Q. B., Sarkar, B., ... & Barman, S. D. (2025). Assessing river water quality for ecological risk in the context of a decaying river in India. *Environmental Science and Pollution Research, 32*(16), 10223–10239. https://doi.org/10.1007/s11356-024-33684-1
Jain, S. K., & Kumar, P. (2014). Environmental flows in India: Towards sustainable water management. *Hydrological Sciences Journal, 59*(3-4), 751–769.
Mahammad, S. (2022). *Palaeochannels and water resource management of Damodar fan-delta, West Bengal*. An unpublished PhD thesis. Aliah University, Kolkata.
Mahammad, S., & Islam, A. (2021). Identification of palaeochannels using optical images and radar data: A study of the Damodar Fan Delta, India. *Arabian Journal of Geosciences, 14*(17), 1–22. https://doi.org/10.1007/s12517-021-07818-5
Mallick, R., Banerjee, A., & Poddar, S. (2025). Sustainable ecosystem management for strengthening climate resilience. In *Sustainable synergy: Harnessing ecosystems for climate resilience* (pp. 219–235). Springer. https://doi.org/10.1007/978-3-031-77957-2_15
Mishra. A., Jena. S.K., Mohanty. S., Sethi. R.R., Kumar. A. (2007). *Generating employment for rural poor through water management interventions: Model design and estimate. Water technology centre for eastern region*. Indian Council of Agricultural Research. http://www.iiwm.res.in/pdf/Bulletin_39.pdf
Palau, A., & Alcázar, J. (2012). The basic flow method for incorporating flow variability in environmental flows. *River Research and Applications, 28*(1), 93–102.
Patra, S. (2008). *Pollution profile of Damodar River in a particular stretch*. Unpublished Report of Master of Technology in Conservation of Rivers and Lakes, Alternate Hydro Energy Centre, Indian Institute of Technology Roorkee. http://shodhbhagirathi.iitr.ac.in:8081/jspui/bitstream/123456789/12088/1/HYDG13946. Pdf
Ramsar Policy Brief on Convention on Wetlands. (2018). *Implementing environmental flows with benefits for society and different wetland ecosystems in river systems*. Retrieved September 25,

2025, from https://www.ramsar.org/sites/default/files/documents/library/rpb4_environmental_flows_e.pdf

Sinha, B., & Rao, P. R. (1985). A study for optimum utilization of the Damodar water resources. *Sadhana, 8*(3), 273–290.

Smakhtin, V., & Anputhas, M. (2006). An assessment of environmental flow requirements of Indian River Basins. *International Water Management Institute's (IWMI's), 107*, 1–42.

The Times of India. (2022). *1,000 check dams to be built in 3 yrs*. Retrieved June 8, 2023, from https://timesofindia.indiatimes.com/city/bhubaneswar/1000-check-dams-to-be-built-in-3-yrs/articleshow/91583777.cms

Tiwary, R. K., & Dhar, B. B. (1994). Environmental pollution from coal mining activities in Damodar river basin, India. *Mine Water and the Environment, 13*(1), 1–10.

WBPCB. (2020). *Revised action plan for rejuvenation of river Damodar Paschim Bardhaman West Bengal*. West Bengal Central Pollution Control Board, River Rejuvenation Committee. https://www.wbpcb.gov.in/writereaddata/files/Action%20Plan%20for%20Riverf%20Damodar.pdf

Yang, S., Zhao, W., Liu, Y., Cherubini, F., Fu, B., & Pereira, P. (2020). Prioritizing sustainable development goals and linking them to ecosystem services: A global expert's knowledge evaluation. *Geography and Sustainability, 1*(4), 321–330. https://doi.org/10.1016/j.geosus.2020.09.004

Zafirah, N., Nurin, N. A., Samsurijan, M. S., Zuknik, M. H., Rafatullah, M., & Syakir, M. I. (2017). Sustainable ecosystem services framework for tropical catchment management: A review. *Sustainability, 9*(4), 546.

Chapter 8
Conclusions

8.1 Major Findings

8.1.1 Damodar Valley Corporation and Quantitative Environmental Flow Alteration

- The environmental flow of the Damodar River has been measured using the hydrological method of Tennant and the flow duration curve (FDC) based on the available period of recorded (POR) discharge data.
- The analysis of the flow and FDC of the river reveals that the flow variability of the river has reduced significantly from the pre-dam to the post-dam period. For example, the annual peak flow has reduced, the frequency of daily high flow has decreased, while daily low flow has increased after the installation of dams and barrages. The flow variability of the river has changed as a result of flow regulation through dams and barrages in the Damodar River system.
- The minimum mean monthly flow in the lean period (December to May) has increased in the post-dam period, i.e., from 1 to 8 m^3/s at the Rhondia site of the river.
- Besides, the mean annual flow has also reduced in the post-dam period, i.e., from 1001.33 to 700.04 m^3/s. The minimum flow requirement was 100.13 m^3/s in the pre-dam period (1934–1957), while it was 70 m^3/s in the post-dam period (1958–2007), which is the prerequisite flow for the river ecosystem. A threshold of 10% mean annual runoff for the river ecosystem was not maintained in the lean season (December to May) in the pre-dam period, whereas it was maintained in the post-dam period at the Rhondia site of the lower Damodar River. This radical change in the flow regimes has altered the channel hydro-morphology, which impacts the eco-geomorphological behaviour of the river.

M. M. Hoque, A. Islam, *Dams, Development and Environmental Flow in the Damodar River Basin*, https://doi.org/10.1007/978-3-032-11161-6_8

- Furthermore, climate change and land use and land cover (LULC) dynamics play an important role in changing the quantitative environmental flow in the river basin.

8.1.2 Damodar Valley Corporation and Qualitative Environmental Flow Alteration

- The present investigation tried to assess the Damodar Valley Corporation (DVC) and qualitative environmental flow alteration in terms of irrigation, drinking and ecological water quality with human health risk. In this regard, the present research used physio-chemical parameters, including major anions and cations.
- The Irrigation Water Quality Index (IWQI) shows that the river water samples vary from unsuitable to very poor, and surface water is relatively less suitable than the bottom water for irrigation use. The river water samples encompass low to medium levels of sodium, salinity, and alkali hazard.
- Most of the physico-chemical parameters' concentration in the water samples lies within their permissible limits according to drinking water standards. The Water Quality Index (WQI) reveals that the river water quality ranges from good to very poor for drinking purposes. The surface river water quality varies from good to poor, while the bottom water quality varies from good to very poor for drinking purposes, i.e., the surface water is more suitable than the bottom water.
- Human health risk analysis portrays that all the values of hazard quotient (HQ) and total hazard index (THI) from both levels of water samples lie under their permissible limit of HQ and THI, i.e.<1. Thus, it may be inferred that all the water samples are free from non-carcinogenic health risks. However, carcinogenic risk analysis depicts that the cancerous risk (CR) values of 50% of water samples from both levels exceeded the permissible limits for adults. About 57.14% of the surface and 50% of bottom water samples exceeded their limits for children, and 21.43% of surface water samples and 28.57% of bottom water exceeded their permissible limits for babies.
- The ecological water quality of the Damodar River is most unacceptable in the pre-monsoon season, while the overall water quality of the river is most unacceptable in the monsoon season. Moreover, a higher eutrophication index of all the stations is recorded, especially during the pre- and post-monsoon periods (dry season), indicating severe eutrophication in the study area. However, bacteriological water quality is most unacceptable in the monsoon in terms of the total coliform (TC) and fecal coliform (FC) concentration, while it is most unacceptable in the premonsoon in terms of the FS concentration. The comprehensive bathing water quality index exhibits the overall unsuitability of the Damodar River for bathing purposes.
- Regarding spatial dynamics, heavy water pollution, as per the overall water quality index (WQI) classification, is found for stretches of the Damodar River down-

stream of the Durgapur barrage. According to the comprehensive pollution index (CPI) classification, 80.52%, 59.09%, and 80.56% of water samples in the pre-monsoon, monsoon and post-monsoon seasons respectively, have slight comprehensive pollution.

- ANOVA depicts that the Damodar River water quality has a significant difference in spatio-temporal dimensions.
- Furthermore, natural and human-driven factors, including climate and LULC changes, play an important role in changing qualitative environmental flows. Emerging contaminants such as pesticides and pharmaceuticals also significantly influence the qualitative environmental flow. Additionally, essential nutrients are being lost downstream due to dam construction.

8.1.3 Dams and Development

- The creation of the Damodar Valley Corporation (DVC) led to the socio-economic development of the inhabitants of the study area. The developing activities, such as controlling floods, supplying water for irrigation, and industrial and municipal usage in the basin area, by the optimum use of available water resources in the basin. It has developed the agricultural sector by providing water for irrigation and improving irrigation facilities in the study area. It is observed that the average crop diversity index (c index) has increased from 0.37 in the pre-dam period (before 1953) to 0.44 in the post-dam period (2022). Similarly, the average cropping intensity has increased from 100% in the pre-dam period to 130.52% in the post-dam period. Crop productivity has also increased from the pre-dam to the post-dam period. For example, the average rice production was 1.8 quintal/*Bigha* in the pre-dam period, while it was 6.57 quintal/*Bigha* in the post-dam period.
- It controls floods and flood probability decreases substantially as several devastating floods occurred in the basin area before the DVC dams. It has developed the industrial sector by supplying water for industrial use. It also supplies water for domestic usage.
- The other activities like strengthening eco-conservation activities, making greenery, constructing roads and embankments, checking soil erosion in the valley area, electricity generation and provision, developing tourism and fish farming in and around the DVC reservoir area, and improving the financial health of the corporation through adopting efficient industrial, commercial, and human resource management practices.
- Besides socio-economic development, dam construction also causes the displacement of local and tribal communities, which is the most significant social impact. They face a livelihood crisis and socio-economic challenges.
- Moreover, LULC dynamics accelerate erosion and increase sediment loads. In the DRB, changes are mainly due to urbanization, agriculture, mining, and industrialization, which significantly increase soil erosion, riverbank erosion,

and sediment load in the river system. LULC changes, specifically, vegetation changes, impact the evapotranspiration status in the basin area. These changes also influence greenhouse gas (GHG) emissions. Additionally, reservoirs can emit GHGs within the basin.

8.1.4 Dams and Environmental Sustainability

- The rapid impact assessment matrix (RIAM) indicates that the positive impact of the DVC project on socio-economic components is comparatively higher in upstream villages, while the negative impact of DVC on physico-chemical (PC) and biological/ecological (BE) components is higher in the downstream villages. However, very weak unsustainability is found for both upstream and downstream villages from the sustainability analysis of the project. The sustainability value is −0.011 for upstream villages and −0.100 for downstream villages. The very weak unsustainability is observed in the Damodar Valley region due to the consequent negative impacts on PC and BE components from the DVC dams.
- Resource management in terms of development activities impacts soil and water quality, leading to habitat loss and biodiversity decline and increasing natural disasters. The water management tools like DVC dams and barrages alter the river's natural flow regime, leading to negative impacts, including riverbed degradation, changing fluvial morphology, and impacting ecosystems. These interventions, along with agricultural and urban-industrial expansion and mining activities, have resulted in the degradation of water quality and caused ecological stress due to pollution releases from these sectors. Moreover, overexploitation of the river water for urban-industrial, agricultural and domestic use also impacts its quality and quantity, leading to consequences on river ecology and biota. These qualitative and quantitative alterations lead to a reduction in riverine habitat and threaten the sustainable development of the basin.
- Anthropogenic activities in terms of development, like industrialisation, agriculture, urbanization, and mining activities impact on groundwater quantity and quality, which affects humans and the ecosystem. Besides, anthropogenically induced climate change may alter precipitation patterns and temperature, which impact the groundwater recharge rate. Though reservoirs creation in the basin area also encourages local groundwater recharge.
- Human activities such as deforestation, hardscaping and other developmental activities impact the quality and quantity of soil water by introducing various contaminants like heavy metals. These activities also increase soil erosion and runoff, and alter the natural soil water balance, changing natural water flow and storage.
- These activities decrease soil productivity due to loss of organic matter, pollution, erosion, and compaction. The anthropogenically driven processes modify soil structure, biological activity and nutrient cycling, which threaten soil fertility, crop production, and ecosystem health. Besides, soil quality degradation dis-

rupts important ecosystem services such as carbon storage, water filtration, and food and raw material supply, affecting livelihoods and leading to socio-economic instability. Moreover, contamination of soil and water and climate change lead to mass migration and conflict over resources. Deforestation, coal, and mineral mining in the basin area lead to excessive soil erosion, resulting in the loss of productive soils and the creation of unproductive wasteland.

- Thus, anthropogenic activities in the context of development lead to several socio-economic developments in the basin, but they also damage the basin's ecosystem, groundwater, soil water, and soil productivity.

8.1.5 Environmental Flow Management

- Environment flow restoration is examined in the context of deteriorating conditions (Class D, not fit for bathing) of the major rivers of India, including the Damodar River as per the findings of the West Bengal Pollution Control Board.
- The geomorphological, hydrological, and ecological health of the Damodar River in the post-dam due to urbanization, industrialization, and agricultural expansion effects are poor, which needs immediate restoration for the revival of the eco-geomorphological behaviour of the Damodar.
- The present study proposes two broad categories of restoration: (1) quantitative techniques aiming to supply river water in a more spatial-temporal continuum, and (2) qualitative techniques aiming to enhance the river water quality for riverine ecology and economy.
- Quantitative flow restorations could be done through—(1) increasing the capacity of the dam reservoir either through desilting in the Mithon or Panchat dams or creating new reservoirs, (2) small-scale interventions like check dams and their maintenance, and (3) local people's knowledge and initiatives. The study established that the mega-scale intervention (new reservoir creation and dredging in the existing reservoir) needs huge costs (about INR 9090 crore required for 10% desilting of Maithon and Panchayet, while INR 6000 crore is required for new reservoir creation at Balpahari) that restricts this measure. Rather, check dams or similar structures and their regular maintenance may be more fruitful in the connection of riverine ecology and economy.
- The field survey indicates that community initiatives and local knowledge might be helpful for flow restorations. The Damodar River Basin (DRB) typically portrays hard rock terrain in the upper part and while alluvial tract in the lower part. Local people of the upper part (e.g., Urwan village) demand the construction dug well for each household and a community tank for rainwater harvesting in the monsoon season. The survey shows that the govt. initiatives coupled with the participation of local people may help the implementation of these projects that may reduce their dependence on the river water in the lean months. On the other hand, for people from the lower alluvial part of the DRB (e.g., Rajarampur vil-

lage), community ponds may be constructed using a participatory approach for the lower dependence on river water and more reliance on rainwater harvesting.
- Qualitative flow restoration could be initiated through the identification of discharging effluent *nallahs*, classification and mapping of industries in the DRB, demarcation of the mining area, and the site suitability mapping of the sewage treatment plant. The identification and locational analysis in the context of the DRB might help the planners to think of further actions so that the less pollutants are added to the Damodar River and its tributaries.

8.2 Confirmation of the Hypothesis

The research work is based on a central hypothesis as mentioned in Chap. 1 is stated below.

> *Dams under the Damodar Valley Corporation, though they play a significant role in developing the regional economy, interrupt the environmental flow for aquatic organisms and human health.*

The hypothesis of the research work has two major aspects: (1) DVC dams play a crucial role in the socio-economic development of the study area, but (2) they interrupt the environmental flow for aquatic organisms and human health. The first aspect related to the socio-economic development of the study area is led by the DVC dams, which have been discussed in Chap. 5 (Table 5.6). It is observed that DVC dams create potential for regional development by serving their beneficiaries as flood control, water usage for irrigation, industrial and domestic, hydropower generation and provision, drought mitigation, recreation, and fisheries. In this context, the role of DVC dams in DRB has been examined in Chap. 5. It is found that agricultural development has occurred in crop diversity, crop intensity, and crop productivity after DVC dams. The ANOVA findings indicate that there is a significant difference in agricultural development between the pre-dam and post-dam periods. Moreover, flood peaks have been moderated in the post-dam period, as examined in Chap. 3.

The second aspect of the hypothesis, i.e., DVC dams interrupt the environmental flow for aquatic organisms and human health, is also discussed in Chaps. 3 and 4 (Tables 4.3). DVC dams interrupt and divert the river's environmental flow. The variability of the river flow is also lost after the dams on the river system. The ANOVA findings indicate that there is a significant difference in the river flow between the pre-dam and post-dam periods. Besides, the water quality of the river for irrigation, drinking, and riverine biota is found in poor conditions. The ANOVA findings indicate that there is a significant statistical difference in water quality in the spatio-temporal dimension (Table 4.3). Thus, the two sub-hypotheses of the present study are confirmed.

8.3 Significance, Limitations, and Way Forward

The work embodied in the present book is the manifestation of the in-depth investigation carried out on the assessment of environmental flow in the context of dams of the DVC. The work has assessed the two vital perspectives of the e. flow—quantitative and qualitative. The moderation of flow fluctuations in the post-dam period becomes significant in the line of flood hydraulics and related socio-economic devastations. However, reduced flow variability does not support habitat suitability, resulting in the reduction of several ecosystem services. This critical perspective on e. flow will certainly help the different stakeholders engaged in e. flow management. The qualitative dimension of e. flow analysis also depicts deteriorating water quality for ecology, agricultural economy (irrigation), and human health risks are really crucial for the water resource manager for sustainable water resource development that reaches the sustainable development goal (SDG 6) i.e. clean water and sanitation. The flow restoration strategies (e.g., increasing dam reservoir capacity building, check dam management, renovation of the pond, construction of new rainwater harvesting structures, identifying discharging effluent *nallahs*, classification and mapping of the industries and mining areas in the DRB) proposed in the present study based on the village level survey fosters integration with the micro level planning that may address the issues of water unavailability, and quality water for life. Therefore, participatory planning involves local people, especially women of the upper part of the basin, especially Jharkhand and the lower part of the basin in West Bengal, state governments, the central government, and local panchayats. Moreover, this work serves as the baseline assessment for future researchers and academicians.

The present study, though carried out with utmost sincerity and in-depth observations, suffers from a few limitations. First, there is a lack of data for water flow and quality analysis, especially in the pre-dam period. This restricted pre-dam and post-dam comparison on several occasions. The second is the lack of secondary data availability related to water quality, especially in the upper part of the basin. This also limited spatial continuity in the hydro-chemical analysis. The third limitation is the unavailability of heavy metals (trace elements) data in the secondary source, especially in the riverbed agricultural field. Fourth, the study used water quality data from different sources—both primary lab tests (from Ramakrishna Mission, Nimpith and National Test House, Salt Lake) and secondary WBPCB data from different spaces and times that may have the problem of data compatibility. Fifth, keystone species are not identified or pattern of path changes for fish migration due to change in flow direction are not mentioned after and before dam construction. Finally, the DVC is the interstate river basin covering about 23,370.98 km^2 area and 18 districts (12 in Jharkhand and 6 in West Bengal state). The people, cultural, and physical landscapes are quite diversified, and a more resonant and diagnostic survey is required before translating the study prescriptions on e. flow restorations at the micro-level.

From the present investigation, several issues crop up for better sustainability and the future of the DRB that might also be helpful to other tropical river basins. For example, the paradigm shifts from 'dam as temple of modern India' to 'decommissioning of DVC dams' needs critical treatment from various stakeholders using a detailed cost-benefit analysis. This may illuminate the future dam dynamics and socio-ecological system exploration from the perspectives of the local, indigenous, and traditional knowledge in combination with modern technology. Methodological innovations like artificial intelligence, machine learning, and robotics may help assist in the future projections on the development of the DVC with different situations like 'dam-controlled development as usual', 'mushrooming of new dams', or 'reduction in dam-control and natural system, restoration'. These issues are really vital in the context of the physically and culturally diversified large basin like the DVC. Future attempts are welcome in these directions for the development of the regional, national, and global tropical south.

Index

M. M. Hoque, A. Islam, *Dams, Development and Environmental Flow in the Damodar River Basin*, https://doi.org/10.1007/978-3-032-11161-6

MIX
Papier aus verantwortungsvollen Quellen
Paper from responsible sources
FSC® C105338

If you have any concerns about our products,
you can contact us on
ProductSafety@springernature.com

In case Publisher is established outside the EU,
the EU authorized representative is:
Springer Nature Customer Service Center GmbH
Europaplatz 3, 69115 Heidelberg, Germany

Printed by Libri Plureos GmbH
in Hamburg, Germany